Studies in Fuzziness and Soft Computing 297

Editor-in-Chief

Prof. Janusz Kacprzyk
Systems Research Institute
Polish Academy of Sciences
ul. Newelska 6
01-447 Warsaw
Poland
E-mail: kacprzyk@ibspan.waw.pl

For further volumes:
http://www.springer.com/series/2941

Dmitri A. Viattchenin

A Heuristic Approach to Possibilistic Clustering: Algorithms and Applications

Dmitri A. Viattchenin
Laboratory of Information Protection
United Institute of Informatics Problems
National Academy of Sciences of Belarus
Minsk
Belarus

ISSN 1434-9922 ISSN 1860-0808 (electronic)
ISBN 978-3-642-44301-5 ISBN 978-3-642-35536-3 (eBook)
DOI 10.1007/978-3-642-35536-3
Springer Heidelberg New York Dordrecht London

Softcover re-print of the Hardcover 1st edition 2013

Printed on acid-free paper

Springer is part of Springer Science+Business Media (www.springer.com)

To my parents

Foreword

Clustering has emerged many decades ago and has formed a fundamental framework supporting various pursuits of data analysis: finding structures in data, determining associations, discovering similarities in time series and spatio-temporal data, building prediction rules, and forming classifiers. Applications of clustering are truly numerous. Just a quick query "clustering" on Google Scholar returns 2,070,000 hits (obviously, one has to take into account some false positives). Fuzzy clustering (coming with 219,000 hits) occupies an important role in the area. This is not surprising by recognizing that fuzzy clustering dwells upon a concept of partial membership - the appealing notion, which is very much in line with the discovery and description of concepts. Here the binary quantification is not in full rapport with the complexities of the phenomena generating the data. Having in mind the principles of Granular Computing, the formal constructs of fuzzy sets, rough sets, sets (intervals) are the semantically sound building blocks used in the algorithms of data analysis.

Fuzzy clustering enjoys a tremendous diversity. There are new concepts, advanced algorithms, and various formalisms exploited. Most of these developments are fueled by a remarkable diversity of applications embracing new areas such as Web mining or social networks. Massive data of quite common these days and some of them are of distributed character generated by various sources. We have been witnessing a variety of pursuits. Among many of them, there are two directions in clustering that have started gaining visibility and importance. Those two endeavors substantially enrich the fundamentals and augment the research agenda. The first one is about clustering with partial supervision or what could be referred to in a more general way as knowledge-based clustering. Typically, clustering is exclusively data-driven and aims at the discovery of a structure in data. Objective function-based clustering algorithms solve optimization problems being guided by a certain performance index and making use of data only. In conjunction to the data themselves, there are some hints-tidbits of knowledge supplied by the user and those pieces are taken into consideration when running the clustering algorithm. There are a number of formats in which the domain knowledge could be articulated. For instance, a subset of labeled data or proximity constraints (where it is said about a closeness/resemblance of some pairs of data) could be considered as a reflection of the available domain knowledge. The proximity can be articulated through the "link" or "not-to-link" dependencies specified between the data. Through knowledge hints the user is actively engaged in the clustering process. In this way the overall process becomes more user-centric. Another important and quite recent trend reported in the literature deals with the granular nature of the data involved in the clustering algorithms. There are applications in which the data are regarded as information granules such as intervals, fuzzy sets, rough sets and alike. These more abstract entities to be clustered call for radical augmentations of the algorithms known to deal with numeric data.

The treatise authored by Dmitri A. Viattchenin offers an interesting and original perspective at possibilistic clustering and uncertain data processing. Along with a classic material on fuzzy and possibilistic clustering, there are new ideas and useful generalizations prudently transformed into an efficient algorithmic environment. It is very much in line with the recent promising development trends identified above. By considering a concept of intuitionistic fuzzy sets, the author recalls generic definitions of the intuitionistic generalization of the clustering method and presents the corresponding clustering methods. A useful approach to the three-way data processing is elaborated on. Various techniques for the interval-valued data processing are considered in detail. A number of crucial topics being central to data analysis are covered: a construction of the set of labeled data for semi-supervised clustering, a methodology for dimensionality reduction of analyzed attribute space, processing of asymmetric data, and a formation of a subset of the most appropriate alternatives for weak fuzzy preference relations. Some directly related design issues such as a rapid prototyping of the Mamdani's fuzzy inference systems are discussed as well.

The book features a well-balanced material and a down-to-the earth exposition. Undoubtedly, the author has provided a comprehensive yet well-focused coverage of the timely topic and offered interesting insights into the area. The material is self-contained and carefully structured. The key concepts, ideas, and algorithms are illustrated by cautiously selected and prudently analyzed examples.

All in all, this volume is a highly welcome contribution to the rapidly growing body of knowledge in contemporary data analysis. The author and the Editor-in-Chief, Professor Janusz Kacprzyk, have to be congratulated on this high quality volume, which broadens our understanding and bolster appreciation of the recent trends in intelligent data analysis and clustering, in particular.

Edmonton and Warsaw Witold Pedrycz

Preface

Fuzzy set theory was proposed by a distinguished American systems and control theorist, Professor Lotfi A. Zadeh, in 1965. Since the appearance of the fundamental Zadeh's paper, fuzzy set theory has been applied to many areas. In particular, many clustering techniques based on fuzzy set theory have been proposed by different researchers and scholars.

Fuzzy clustering methods and possibilistic clustering methods are very popular and useful techniques for data processing. These methods belong the most frequently pursued activities of researchers, engineers, managers, programmers, and other professionals. Fuzzy clustering offers special methods and procedures to solve many practical problems. Thousands of successful applications of fuzzy clustering and possibilistic clustering methods are described in a ultitude of publications and these applications confirm a real applicability of those tools and techniques in many areas of science, engineering, defence, medicine, and business and economics, to name a few.

The present book is meant to present and analyze a new approach to possibilistic clustering, and to show some interesting and relevant applications of the proposed methods. The essence of the proposed approach is that the sought clustering structure of the set of objects is formed based directly on the formal definition of a fuzzy cluster and possibilistic memberships are also determined directly from the values of the pairwise similarity comparisons of objects.

All methods developed by the author and described in the book are illustrated with examples. Of course, many problems discussed in the book require more detailed a study. However, the book in its resent form, with an adequate degree of rigour and a constructive approach, can be interesting and useful for many people in the areas of computer science, artificial intelligence, data analysis, data mining, etc. Moreover, the book is meant for analysts, engineers, scientists, scholars, students and post-graduate students who deak with fuzzy clustering and its applications. I hope that the present book will be a valuable reference work for both theorists and practiciners.

The first chapter introduces the basic concepts considered in the book. In particular, the basic concepts of fuzzy sets theory, possibility theory, some basic fuzzy clustering and possibilistic clustering methods and techniques are biefly considered.

The second chapter focuses on heuristic algorithms of possibilistic clustering. So, basic concepts of the proposed approach to clustering and its corresponding algorithms are described. Results of some numerical experiments are also given.

The third chapter includes the description of techniques for uncertain data processing using the clustering methods introduced. In particular, basic definitions of aintuitionistic fuzzy generalization of the clustering method and the corresponding clustering procedure are given, some concepts and techniques of fuzzy data processing are introduced, the technique of three-way data processing is described, and techniques of interval-valued data processing are considered in detail.

The fourth chapter presents an outline of techniques for the application of the proposed clustering approach to solve some applied problems in data analysis. To be more specific, the methods for constructing the set of labeled objects for the semi-supervised clustering algorithm, for the reduction of dimensionality of the analyzed attribute space, for the asymmetric data processing, for constructing a subset of the most appropriate alternatives for a set of weak fuzzy preference relations defined on an universe of alternatives, are described in detail. Moreover, a method of rapid prototyping for the Mamdani fuzzy inference systems is shown.

The present book is the result of research done at the Systems Research Institute, Polish Academy of Sciences and at the Ostfalia University of Applied Sciences during the last seven years. I would like to thank Director of the Systems Research Institute, the Polish Academy of Sciences for the possibility of that reserach and the Mianowski Fund for financial support. A partial support by the German Academic Exchange Service is also acknowledged. I am grateful to Prof. Janusz Kacprzyk for the invitation to publish this book in his series, for his supervision, his support, fruitful discussions and useful remarks during the manuscript preparation. I would like to thank Dr. Jan W. Owsiński, Professor Eulalia Szmidt, Professor Walenty Ostasiewicz, Professor Frank Klawonn, Professor Reinhard Viertl, Professor Vladik Kreinovich, Professor Mika Sato-Ilic and Professor Zeshui Xu for their interest in my investigations, collaboration and support. Special thanks are addressed to Mr. Aliaksandr Damaratski, Mr. Pavel Savyhin, and Mr. Dzmitry Novikau for developing the experimental software. Finally, I express my deep gratitude to the editorial staff of Springer for their wonderful co-operation.

Warsaw, Poland Dmitri A. Viattchenin

Contents

Chapter 1
Introduction

To begin with, *cluster analysis* is a structural approach to solving the problem of object classification without training samples. Clustering methods aim at partitioning of a set of objects into subsets, called *clusters*, so that the objects belonging to the same cluster aare as similar as possible, and on the other hand, the objects belonging to different clusters are as dissimilar as possible. The resulting data partitioning improves data understanding and reveals its internal structure. Clustering methods are also called automatic classification methods and numerical taxonomy methods. Heuristic, hierarchical, optimization and approximation methods are main approaches to cluster analysis.

Since the fundamental Zadeh's paper [169], fuzzy sets theory has been applied to many areas such as decision-making, control, classification, learning, etc.. The idea of a fuzzy approach to classification was outlined by Bellmann, Kalaba and Zadeh [8]. Fuzzy sets theory makes it possible to model partial belongingness to a cluster which is described by a membership function. Thus, the representativeness of each object for a single cluster is determined during the analysis. Fuzzy clustering methods have been effectively applied in image processing, data analysis, symbol recognition, modeling, to name a few. Heuristic methods of fuzzy clustering, hierarchical methods of fuzzy clustering and optimization methods of fuzzy clustering were proposed by different researchers. Moreover, a possibilistic approach to clustering was proposed in Krishnapuram and Keller[66] and then further developed in other publications.

In this chapter some basic definitions of fuzzy sets theory are given as well as some existing fuzzy clustering methods and possibility clustering techniques are considered.

1.1 Fundamentals of Fuzzy Sets Theory

The first subsection of the section provides a brief description of basic definitionsof fuzzy sets theory and possibility theory. Fuzzy relations are considered in the second subsection. In the third subsection fuzzy numbers are described.

D.A. Viattchenin: *A Heuristic Approach to Possibilistic Clustering*, Studfuzz 297, pp. 1 – 58.
DOI: 10.1007/978-3-642-35536-3_1

1.1.1 Fuzzy Sets and Possibility Theory

Let us remind some basic definitions of fuzzy sets theory which will be used in further considerations. So, a definition of a fuzzy set must be considered at the first place.

Definition 1.1. *Let X denote some universe of discourse. Then a fuzzy set A in X is a mapping $\mu_A : X \to [0,1]$, which assigns to each object $x \in X$ a membership degree $\mu_A(x)$.*

In other words, a fuzzy set A on X is a set of ordered pairs $A = \{x, \mu_A(x)\}$, $x \in X$. If the degree $\mu_A(x)$ is close to 1, then the membership degree of x to A becomes high. Otherwise, the degree $\mu_A(x) = 0$ represents full non-membership in a fuzzy set A. If $X = \{x_1, \ldots, x_n\}$ is finite, then a fuzzy set A can be expressed as $A = \sum_{i=1}^{n} \mu_i(x_i)/x_i = \mu_1(x_1)/x_1 + \ldots + \mu_n(x_n)/x_n$, where the plus sign plays the role of the union. If X is infinite, then the notation $A = \int_X \mu_A(x)/x$ is be used. The set of all fuzzy sets of X is denoted by $[0,1]^X$.

An explanation is here needed. For example, the continuous membership function of a fuzzy set $A^1 = \int_0^1 \left(1/1 + x^2\right)/x$ is shown as a continuous curve in Figure 1.1, where $x \in X$ and $X = [0, 1]$ is the universe. On the otherhand, the discrete membership function $\mu_{A^2}(x_i) = 1/1 + x_i^2$, $x_i \in X$, of a fuzzy set A^2 is shown by • in the Figure 1.1, where $X = \{ x_1 = 0.0, x_2 = 0.2, x_3 = 0.4, x_4 = 0.6, x_5 = 0.8, x_6 = 1.0\}$ is the universe and $\mu_A(x)$ is a generic notation for the membership functions of fuzzy sets A^1 and A^2 in Figure 1.1.

Definition 1.2. *Let X besome universe of discourse and A bea fuzzy set in X with $\mu_A(x)$ being its membership function. The support $Supp(A)$ of the fuzzy set A is the set of elements of X fr which $\mu_A(x) > 0$.*

In other words, $Supp(A) = \{x \in X \mid \mu_A(x) > 0\}$. For example, Figure 1.1 shows that all points x_i of the universe $X = \{ x_1, \ldots, x_6\}$ are elements of $Supp(A^2)$ of the fuzzy set A^2 with the discrete membership function $\mu_{A^2}(x_i)$, and all points $x \in X = [0,1]$ are elements of $Supp(A^1)$ for the fuzzy set A^1 withthe continuous membership function $\mu_{A^1}(x)$.

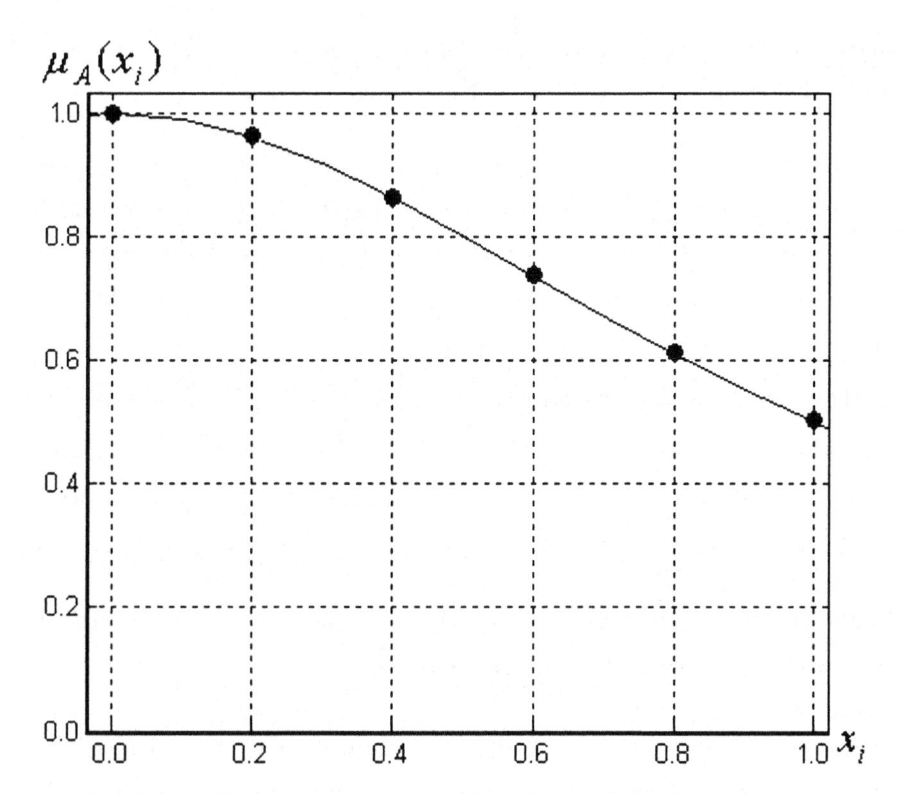

Fig. 1.1 Continuous and discrete membership functions of fuzzy sets

Definition 1.3. *Let A be a fuzzy set in some universe X with $\mu_A(x)$ being its membership function. The height of A is defined by the expression*

$$h(A) = \sup_{x \in X} \mu_A(x). \tag{1.1}$$

If the condition $h(A) = 1$ is met, then a fuzzy set A is normal. Otherwise, if $h(A) < 1$, then A is subnormal. For example, in Figure 1.1 the fuzzy sets A^1 and A^2 are normal.

Definition 1.4. *The core $Core(A)$ of a fuzzy set A in some universe X is the set of elements in X for which $\mu_A(x) = 1$.*

In Figure 1.1, the element $x = 0.0$ is a unique element of core for both the fuzzy sets A^1 and A^2. So, a core of a subnormal fuzzy set is an empty set.

Definition 1.5. *The α-level set or the α-cut A_α of the fuzzy set A is defined as the set of elements in X for which $\mu_A(x) \geq \alpha$, where $0 < \alpha \leq 1$.*

So, the α-level A_α can be defined as $A_\alpha = \{x \in X \mid \mu_A(x) \geq \alpha\}$. If the universe $X = \{x_1, \ldots, x_n\}$ is finite, then all values $\alpha \in (0,1]$ can be written as an ordered sequence $0 < \ldots < \alpha_\ell < \ldots \leq 1$. In Figure 1.1, the α-level set $A^2_{\alpha_\ell}$ of the fuzzy set A^2 for $\alpha_\ell = 0.8$ is the set of elements $A^2_{\alpha_\ell = 0.8} = \{ x_1 = 0.0, x_2 = 0.2, x_3 = 0.4\}$.

The strong α-level set $A_{\overline{\alpha}_\ell}$ of a fuzzy set A is defined as $A_{\alpha_\ell} = \{x \in X \mid \mu_A(x) > \alpha_\ell\}$, $\alpha_\ell \in (0,1]$. That is why the support *Supp*(*A*) of the fuzzy set A in X can be defined as the strong α-level set $A_{\overline{\alpha}_\ell}$ of the fuzzy set A for $\alpha_\ell = 0$.

Definition 1.6. *The α-level fuzzy set $A_{(\alpha_\ell)}$ of a fuzzy set A is defined as $A_{(\alpha_\ell)} = \{(x, \mu_A(x)) \mid x \in A_{\alpha_\ell}\}$, where $0 < \alpha_\ell \leq 1$.*

The strong α-level fuzzy set $A_{(\overline{\alpha}_\ell)}$ of a fuzzy set A is defined as $A_{(\overline{\alpha}_\ell)} = \{(x, \mu_A(x)) \mid x \in A_{\overline{\alpha}_\ell}\}$. In Figure 1.1, the strong α-level fuzzy set $A^2_{(\overline{\alpha}_\ell)}$ of the fuzzy set A^2 for $\alpha_\ell = 0.8$ is the fuzzy set $A^2_{(\overline{\alpha}_\ell)} = \sum_{i=1}^{3} \mu_i(x_i)/x_i = 1.00/0.0 + 0.96/0.2 + 0.86/0.4$. The concept of the α-level fuzzy set was introduced by Radecki [94].

Some basic operations for fuzzy sets must be also considered. Let A and B are fuzzy sets in some infinite universe X with corresponding continuous membership functions $\mu_A(x)$ and $\mu_B(x)$.

Definition 1.7. *If th condition $\mu_A(x) \leq \mu_A(x)$, $\forall x \in X$ is met then the fuzzy set A is included in (is a subset of) the fuzzy set B, $A \subseteq B$.*

If the condition $\mu_A(x) < \mu_B(x)$, $\forall x \in X$ is met, then the strong inclusion is denoted by $A \subset B$.

Definition 1.8. *If the condition $\mu_A(x) = \mu_A(x)$, $\forall x \in X$ is met, then the fuzzy set A is equal to the fuzzy set B, $A = B$.*

Let A is a fuzzy set in some infinite universe X and A_{α_1}, A_{α_2} are its α-levels sets for the corresponding values $\alpha_1, \alpha_2 \in (0,1]$. Obviously, the condition

$$\alpha_2 \geq \alpha_1 \Rightarrow A_{\alpha_2} \subseteq A_{\alpha_1}, \tag{1.2}$$

is met for the α-levels sets of A. So, a very important proposition was formulated and proved, for example, by Kaufmann [60].

Theorem 1.1. *Let A be a fuzzy set in some universe X with $\mu_A(x)$ being its membership function and $\{A_{\alpha_\ell} \mid \alpha_\ell \in [0,1]\}$ be a family of α-levels sets of the fuzzy set A for the ordered sequence $0 < \ldots < \alpha_\ell < \ldots \leq 1$. The fuzzy set A can be represented as follows:*

$$\mu_A(x) = \max_{\alpha_\ell}(\alpha_\ell \cdot \mu_{A_{\alpha_\ell}}(x)), \tag{1.3}$$

where

$$\mu_{A_{\alpha_\ell}}(x) = \begin{cases} 1, \text{ if } \mu_A(x) \geq \alpha_\ell \\ 0, \text{ if } \mu_A(x) < \alpha_\ell \end{cases}, \tag{1.4}$$

for all $\alpha_\ell \in (0,1]$.

The operation of complementation is equivalent to negation (the connective "not"), and this very important operation also must be defined.

Definition 1.9. *The complement of the fuzzy set A, denoted by $\overline{A}$, is defined as*

$$\overline{A} = \int_X (1 - \mu_A(x))/x. \tag{1.5}$$

Important concepts related to the fuzzy sets are those of a T-norm (triangular norm) and S-norm (triangular conorm). The T-norms and S-norms are defined as a class of intersection or aggregation operators, for fuzzy sets in our context. Let us consider the T-norms and S-norms for some infinite universe of discourse X.

For some three fuzzy sets A, B and C in X with their membership functions $\mu_A(x)$, $\mu_B(x)$ and $\mu_C(x)$, $\forall x \in X$, the T-norms the function $\mathrm{T}: [0,1] \times [0,1] \rightarrow [0,1]$ satisfying the following conditions:

- Commutativity:

$$\mathrm{T}(\mu_A(x), \mu_B(x)) = \mathrm{T}(\mu_B(x), \mu_A(x)), \tag{1.6}$$

- Associativity:

$$\mathrm{T}(\mu_A(x),\mathrm{T}(\mu_B(x),\mu_C(x)))=\mathrm{T}(\mathrm{T}(\mu_A(x),\mu_B(x)),\mu_C(x)), \quad (1.7)$$

- Monotonicity:

$$\mu_B(x)\leq\mu_C(x)\Rightarrow\mathrm{T}(\mu_A(x),\mu_B(x))\leq\mathrm{T}(\mu_A(x),\mu_C(x)), \quad (1.8)$$

- Neutral element 1:

$$\mathrm{T}(\mu_A(x),1)=\mathrm{T}(1,\mu_A(x))=\mu_A(x). \quad (1.9)$$

There are many parameterized families of T-norms which were considered, for example, in [175]. In practical applications, the most widely used T-norms are [33]:

- Minimum operation:

$$\mathrm{T}_1(\mu_A(x),\mu_B(x))=\min(\mu_A(x),\mu_B(x)), \quad (1.10)$$

- Algebraic product:

$$\mathrm{T}_2(\mu_A(x),\mu_B(x))=\mu_A(x)\cdot\mu_B(x), \quad (1.11)$$

- Bounded product:

$$\mathrm{T}_3(\mu_A(x),\mu_B(x))=\max(0,\mu_A(x)+\mu_B(x)-1). \quad (1.12)$$

The dual notion to the T-norm is the S-norm. Its neutral element is 0 instead of 1, and all other conditions remain unchanged. So, for three fuzzy sets A, B and C in X with their membership functions $\mu_A(x)$, $\mu_B(x)$ and $\mu_C(x)$, $\forall x\in X$, the S-norms the function $\mathrm{S}:[0,1]\times[0,1]\rightarrow[0,1]$ satisfying the following conditions:

- Commutativity:

$$\mathrm{S}(\mu_A(x),\mu_B(x))=\mathrm{S}(\mu_B(x),\mu_A(x)), \quad (1.13)$$

- Associativity:

$$\mathrm{S}(\mu_A(x),\mathrm{S}(\mu_B(x),\mu_C(x)))=\mathrm{S}(\mathrm{S}(\mu_A(x),\mu_B(x)),\mu_C(x)), \quad (1.14)$$

- Monotonicity:

$$\mu_B(x) \le \mu_C(x) \Rightarrow \mathrm{S}(\mu_A(x), \mu_B(x)) \le \mathrm{S}(\mu_A(x), \mu_C(x)), \tag{1.15}$$

- Neutral element 0:

$$\mathrm{S}(\mu_A(x), 0) = \mathrm{S}(0, \mu_A(x)) = \mu_A(x). \tag{1.16}$$

If T is a T-norm, then $\mathrm{S}(\mu_A(x_i), \mu_B(x_i)) = 1 - \mathrm{T}(1 - \mu_A(x_i), 1 - \mu_B(x_i))$ is an S-norm, and vice versa. For example, the most widely used S-norms are [33]:

- Maximum operation:

$$\mathrm{S}_1(\mu_A(x), \mu_B(x)) = \max(\mu_A(x), \mu_B(x)), \tag{1.17}$$

- Probabilistic sum:

$$\mathrm{S}_2(\mu_A(x), \mu_B(x)) = \mu_A(x) + \mu_B(x) - \mu_A(x) \cdot \mu_B(x), \tag{1.18}$$

- Bounded sum:

$$\mathrm{S}_3(\mu_A(x), \mu_B(x)) = \min(1, \mu_A(x) + \mu_B(x)). \tag{1.19}$$

Therefore, the T-norms T_q and S-norms S_q, $q \in \{1, 2, 3\}$, are dual pairs. Note that no T-norm can yield greater values than the minimum operation T_1 and no S-norm can yield smaller values than the maximum operation S_1. The T-norms and S-norms can be formulated immediately for some finite universe of discourse X.

Some other operations for fuzzy sets, such as the product, concentration and dilation of fuzzy sets are considered, for example, by Friedman and Kandel [36].

Note that some extensions for fuzzy sets were proposed. In particular, type-two fuzzy sets were proposed in [171] and investigated in other publications. On the other hand, in [4] Atanassov extended Zadeh's fuzzy set to a more general form which was called intuitionistic fuzzy set. The intuitionistic fuzzy set is characterized by two functions expressing the degree of membership and the degree of non-memberships, respectively. Type-two fuzzy sets and intuitionistic fuzzy sets will be used in our further considerations. Moreover, fuzzy sets theory is a basis for possibility theory which was proposed by Zadeh [172] and then developed in [33].

The fundamental concept of possibility theory is the possibility distribution, which assigns to each element x_i, $i = 1, \ldots, n$ in a set $X = \{x_1, \ldots, x_p\}$ a degree of possibility $\pi(x_i) \in [0,1]$ of being the correct description of a state of affairs. By convention, $\pi(x_i) = 1$ means that it is fully possible; $\pi(x_i) = 0$ means that x_i is impossible. A possibility distribution π is said to be normalized if there exists at least one $x_i \in X$ which is totally possible. So, a condition $\sup_{x_i \in X}\{\pi(x_i)\} = 1$ is met. In the case of a sub-normalized π, a value which is defined by

$$I(\pi) = 1 - \sup_{x_i \in X}\{\pi(x_i)\}, \tag{1.20}$$

is called the inconsistency degree of π. So, for a normalized π, $\sup_{x_i \in X}\{\pi(x_i)\} = 1$, hence $I(\pi) = 0$.

From a possibility distribution, two dual measures can be derived: the possibility and necessity measures. Given a possibility distribution π on the universe X, the corresponding possibility and necessity measures of any event $O \subseteq 2^X$ are, respectively, determined by the formulae:

$$\Pi(O) = \sup_{x_i \in O} \pi(x_i)$$

and

$$\mathrm{N}(O) = \inf_{x_i \notin O} (1 - \pi(x_i)) .$$

A possibility measure $\Pi(O)$ expresses at which level O is consistent with our knowledge represented by π, while $\mathrm{N}(O) = 1 - \Pi(\overline{O})$ expesses at which level O is certainly implied by our knowledge.

Let A be a fuzzy set in some universe X with $\mu_A(x)$ being its membership function. Let $[0,1]^X$ be the set of all fuzzy sets of the universe X. So, $\forall \Pi$, $\exists A \in [0,1]^X$, $\forall x_i \in X$, $\Pi(\{x_i\}) = \pi(x_i) = \mu_A(x_i)$. On the other hand, $\forall A \in [0,1]^X$, $\exists \Pi$, $\forall x_i \in X$, $\Pi(\{x_i\}) = \Pi(x_i) = \mu_A(x_i)$. These conditions are considered by Dubois and Prade [33]. That is why a possibility distribution can be interpreted as a membership function of some fuzzy set in the universe X. Note that $\Pi(X) = \sup_{x_i \in X} \mu_A(x_i)$ and the value is the height of the fuzzy set A in the universe X.

Basic concepts of possibility theory are very important for our further consideration of possibility clustering methods. Moreover, all presented arguments can be immediately extended to any infinite universe X.

1.1.2 Fuzzy Relations

Fuzzy relations play an important role in clustering. Moreover, the initial data for clustering can be presented as a set of fuzzy relations. That is why fuzzy relations should be considered in detail.

Definition 1.10. *Let X and Y denote universes of discourse. Then a fuzzy set R of a Cartesian product $X \times Y$ is called a fuzzy relation between the universes X and Y, and $\mu_R(x, y)$, $\forall x \in X, \forall y \in Y$ is the membership function of the binary fuzzy relation R.*

So, if X and Y are infinite sets, then a fuzzy relation R can be expressed as $R = \int_{X \times Y} \mu_R(x, y)/(x, y)$. If the universes $X = \{x_1, \ldots, x_n\}$ and $Y = \{y_1, \ldots, y_p\}$ are finite, then the notation $\mu_R(x_i, y_j)$, $x_i \in X$, $y_j \in Y$ will be used for the fuzzy relation R. Let us remind concepts ofprojections of a fuzzy relation which were considered by Kaufmann [60].

Definition 1.11. *Let X and Y be some finite universesand R be a binary fuzzy relation on $X \times Y$ with $\mu_R(x_i, y_j)$ being its membership function. The first projection of the fuzzy relation R denoted by $\mathrm{Proj}_1(R)$, is defined by*

$$\mathrm{Proj}_1(R) = \max_{y_j} \mu_R(x_i, y_j), \; \forall \, x_i \in X, \; y_j \in Y. \tag{1.21}$$

Definition 1.12. *The second projection of the fuzzy relation R on $X \times Y$ denoted by $\mathrm{Proj}_2(R)$, is defined by*

$$\mathrm{Proj}_2(R) = \max_{x_i} \mu_R(x_i, y_j), \; \forall \, x_i \in X, \; y_j \in Y. \tag{1.22}$$

Definition 1.13. *The global projection of the binary fuzzy relation R on $X \times Y$ denoted by $\mathrm{Proj}(R)$, is defined by*

$$\mathrm{Proj}(R) = \max_{x_i} \max_{y_j} \mu_R(x_i, y_j) = \max_{y_j} \max_{x_i} \mu_R(x_i, y_j), \; \forall \, x_i \in X, \; y_j \in Y. \tag{1.23}$$

If the condition $\mathrm{Proj}(R)=1$ is met, then a fuzzy relation R on $X \times Y$ is the normal fuzzy relation. Otherwise, if $\mathrm{Proj}(R)<1$, then a fuzzy relation R is the subnormal fuzzy relation. The very important concepts of projections of a fuzzy relation should be explained on an example.

Let us assume that $X=\{x_1,x_2\}=\{1,2\}$ and $Y=\{y_1,y_2,y_3\}=\{0,0.1,0.2\}$ are universes of discourse. The fuzzy relation R on $X \times Y$ with the membership function $\mu_R(x_i,y_j)$ is presented in Table 1.1 and this fuzzy relation can be interpreted as "x is considerably larger than y".

Table 1.1 Fuzzy relation R "x is considerably larger than y" and its projections

R	y_1	y_2	y_3	$\mathrm{Proj}_1(R)$
x_1	0.5	0.4	0.4	0.5
x_2	1.0	0.9	0.8	1.0
$\mathrm{Proj}_2(R)$	1.0	0.9	0.8	1.0
				$\mathrm{Proj}(R)$

Obviously, the fuzzy relation R is the normal fuzzy relation because $\mu_R(x_2,y_1)=1$ and $\mathrm{Proj}(R)=1$. So, if the condition $\mu_R(x_i,y_j)=1$ is met for some pair $(x_i,y_j)\in X \times Y$, then the fuzzy relation R on $X \times Y$ is the normal fuzzy relation.

From the possibilistic point of view, if R is some binary fuzzy relation on $X \times Y$, then $\mu_R(x_i,y_j)=\pi(x_i,y_j)$, $x_i \in X$, $y_j \in Y$, and the first projection $\mathrm{Proj}_1(R)$ of the fuzzy relation R can be expressed by the following possibility distribution:

$$\pi_1(x_i)=\Pi(\{x_i\}\times Y)=\sup_{y_j}\pi(x_i,y_j). \tag{1.24}$$

as demonstrated by Dubois and Prade in [33].

If $X=Y$, then the fuzzy relation R is the fuzzy relation on the universe X. Fuzzy relations on the finite universe $X=\{x_1,\ldots,x_n\}$ will be considered in the further exposition.

Definition 1.14. *Let* X *be the universe and* R *be a fuzzy relation on* X *with* $\mu_R(x_i, x_j)$ *being its membership function. The reverse fuzzy relation* R^{-1} *on* X *is defined by the membership function* $\mu_{R^{-1}}(x_i, x_j) = \mu_R(x_j, x_i)$, $\forall x_i, x_j \in X$.

Definition 1.15. *Let* X *be some finite universe and* R *be a fuzzy relation on* X *with* $\mu_R(x_i, x_j)$ *being its membership function. The support* $Supp(R)$ *of the fuzzy relation* R *is the set of pairs in* $(x_i, x_j) \in X \times X$ *at which* $\mu_R(x_i, x_j) > 0$.

Definition 1.16. *The* α*-level set* R_{α_ℓ} *of the fuzzy relation* R *is defined as the set of ordered pairs of objects* $(x_i, x_j) \in X \times X$ *at which* $\mu_R(x_i, x_j) \geq \alpha_\ell$, *where* $0 \leq \alpha_\ell \leq 1$.

Thus, the α-level R_{α_ℓ} of the fuzzy relation R can be defined as $R_{\alpha_\ell} = \{(x_i, x_j) \in X \times X \mid \mu_R(x_i, x_j) \geq \alpha_\ell\}$. Moreover, the support $Supp(R)$ of the fuzzy relation R on X can be defined as the strong α-level set $R_{\overline{\alpha}_\ell}$ of the fuzzy relation R for $\alpha_\ell = 0$, that is $Supp(R) = \{(x_i, x_j) \in X \times X \mid \mu_R(x_i, x_j) > 0\}$. The following definition is very important and will be used in he further considerations.

Definition 1.17. *Let* X *be a finite universe and* R *be a fuzzy relation on* X *with* $\mu_R(x_i, x_j)$ *being its membership function. The* α*-level fuzzy relation* $R_{(\alpha_\ell)}$ *of the fuzzy relation* R *on* X *is defined as* $R_{(\alpha_\ell)} = \{\langle (x_i, x_j) \in R_{\alpha_\ell}, \mu_{R_{(\alpha_\ell)}}(x_i, x_j) = \mu_R(x_i, x_j) \rangle\}$, *where* $0 < \alpha_\ell \leq 1$.

So, the membership function $\mu_{R_{(\alpha_\ell)}}(x_i, x_j)$ of the α-level fuzzy relation $R_{(\alpha_\ell)}$ can be defined as follows:

$$\mu_{R_{(\alpha_\ell)}}(x_i, x_j) = \begin{cases} \mu_R(x_i, x_j), & \text{if } \mu_R(x_i, x_j) \geq \alpha_\ell \\ 0, & \text{if } \mu_R(x_i, x_j) < \alpha_\ell \end{cases}, \tag{1.25}$$

where $\alpha_\ell \in (0,1]$. The definition of the α-level fuzzy relation $R_{(\alpha_\ell)}$ was proposed in [117].

Let us consider basic operations for fuzzy relations. The operation of complementation of a fuzzy relation R on X is similar to the operation for fuzzy sets.

Definition 1.18. *The complement of the fuzzy relation* R *on* X *is a fuzzy relation* $\overline{R}$ *on* X *defined by the membership function*

$$\mu_{\overline{R}}(x_i, x_j) = 1 - \mu_R(x_i, x_j), \tag{1.26}$$

for all $x_i, x_j \in X$.

Let R and Q are fuzzy relations, defined on some finite universe X with corresponding membership functions $\mu_R(x_i, x_j)$ and $\mu_Q(x_i, x_j)$. Now, two definitions of relationships between the fuzzy relations R and Q can be defined.

Definition 1.19. *If the condition* $\mu_R(x_i, x_j) \leq \mu_Q(x_i, x_j)$, $\forall (x_i, x_j) \in X \times X$ *is met, then the fuzzy relation* R *is included in the fuzzy relation* Q, $R \subseteq Q$.

Definition 1.20. *If the condition* $\mu_R(x_i, x_j) = \mu_Q(x_i, x_j)$, $\forall (x_i, x_j) \in X \times X$ *is met, then the fuzzy relation* R *is equal to the fuzzy relation* Q, $R = Q$.

Moreover, the operations which were defined for the fuzzy sets can be also extended for the fuzzy relations (whch are fuzzy sets defined in the Cartesian products). In particular, the T-norms and S-norms can be defined as a class of intersection and aggregation operators for fuzzy relations. However, the intersection and theunion of fuzzy relations R and Q will be defined here as follows.

Definition 1.21. *For two fuzzy relations* R *and* Q, *defined on* X, *their intersection* $R \cap Q$ *is defined by the membership function*

$$\mu_{R \cap Q}(x_i, x_j) = \min\left(\mu_R(x_i, x_j), \mu_Q(x_i, x_j)\right). \tag{1.27}$$

Definition 1.22. *For two fuzzy relations* R *and* Q, *defined on* X, *their union* $R \cup Q$ *is defined by the membership function*

$$\mu_{R \cup Q}(x_i, x_j) = \max\left(\mu_R(x_i, x_j), \mu_Q(x_i, x_j)\right). \tag{1.28}$$

Obviously, the intersection (1.27) and the union (1.28) are similar to the minimum operation (1.10) and the maximum operation (1.17) for the fuzzy sets. Moreover, the intersection (1.27) and the union (1.28) satisfied to the following conditions:

- Commutativity:

$$R \cap Q = Q \cap R, \tag{1.29}$$

$$R \cup Q = Q \cup R. \tag{1.30}$$

- Associativity:

$$R \cap (Q \cap S) = (R \cap Q) \cap S, \tag{1.31}$$

$$R \cup (Q \cup S) = (R \cup Q) \cup S. \tag{1.32}$$

Definition 1.23. *Let R and Q be two fuzzy relations, defined on X. Their (max-min) composition is defined to be the new fuzzy relation $R \circ Q$ on X with the membership function*

$$\mu_{R \circ Q}(x_i, x_k) = \max_{x_j \in X}\left(\min(\mu_R(x_i, x_j), \mu_Q(x_j, x_k))\right), \tag{1.33}$$

for all $x_i, x_j, x_k \in X$.

Let R, Q and S are three fuzzy relations, defined on some finite universe X. The (max-min) composition is associative:

$$R \circ (Q \circ S) = (R \circ Q) \circ S. \tag{1.34}$$

Let us consider some important properties of the fuzzy relations which will be useful in our further exposition:

- Reflexivity:

$$\mu_R(x_i, x_i) = 1, \forall x_i \in X, \tag{1.35}$$

- Strong reflexivity:

$$\mu_R(x_i, x_i) = 1, \mu_R(x_i, x_j) < 1, \ \forall x_i, x_j \in X, \ x_i \neq x_j, \tag{1.36}$$

- Weak reflexivity:

$$\mu_R(x_i, x_j) \leq \mu_R(x_i, x_i), \forall x_i, x_j \in X, \tag{1.37}$$

- Antireflexivity:

$$\mu_R(x_i, x_i) = 0, \forall x_i \in X, \tag{1.38}$$

- Strong antireflexivity:

$$\mu_R(x_i, x_i) = 0, \mu_R(x_i, x_j) > 0, \ \forall x_i, x_j \in X, \ x_i \neq x_j, \tag{1.39}$$

- Weak antireflexivity:

$$\mu_R(x_i, x_j) \geq \mu_R(x_i, x_i), \forall x_i, x_j \in X, \tag{1.40}$$

- Symmetry:

$$\mu_R(x_i, x_j) = \mu_R(x_j, x_i), \forall x_i, x_j \in X, \tag{1.41}$$

- Antisymmetry:

$$\mu_R(x_i, x_j) \neq \mu_R(x_j, x_i) \text{ or } \mu_R(x_i, x_j) = \mu_R(x_j, x_i) = 0, \forall x_i, x_j \in X, \ x_i \neq x_j, \tag{1.42}$$

- Perfect antisymmetry:

$$\mu_R(x_i, x_j) > 0 \Rightarrow \mu_R(x_j, x_i) = 0, \forall x_i, x_j \in X, \ x_i \neq x_j, \tag{1.43}$$

- (Max-min) transitivity:

$$\mu_R(x_i, x_k) \geq \max_{x_j \in X}\left(\min(\mu_R(x_i, x_j), \mu_R(x_j, x_k))\right), \forall x_i, x_j, x_k \in X, \tag{1.44}$$

- (Min-max) transitivity:

$$\mu_R(x_i, x_k) \leq \min_{x_j \in X}\left(\max(\mu_R(x_i, x_j), \mu_R(x_j, x_k))\right), \forall x_i, x_j, x_k \in X. \tag{1.45}$$

Some properties can be defined in other terms. For example, the symmetry condition can be defined by the expression

$$\mu_R(x_i, x_j) = \mu_{R^{-1}}(x_i, x_j), \forall x_i, x_j \in X. \tag{1.46}$$

On the other hand, the (max-min) transitivity condition can be rewritten through the (max-min) composition as follows:

$$\mu_{R \circ R}(x_i, x_j) \leq \mu_R(x_i, x_j), \forall x_i, x_j \in X. \tag{1.47}$$

Definition 1.24. *Let* $X = \{x_1, \ldots, x_n\}$ *be a finite universe,* $\mathrm{card}(X) = n$, *and* R *be a fuzzy intransitive relation on* X *with* $\mu_R(x_i, x_j)$ *being its membership function. The fuzzy binary relation* $\widetilde{R}$ *is the(max-min) transitive closure of the fuzzy relation* R *and* $\widetilde{R}$ *is defined by the expression*

$$\widetilde{R} = R^1 \cup R^2 \cup \ldots \cup R^n, \tag{1.48}$$

where

$$R^1 = R, \; R^g = R^{g-1} \circ R, \; g = 2, \ldots, n, \tag{1.49}$$

and the union of two different fuzzy relations is defined by the expression (1.28), and the (max-min) composition of two fuzzy relations is defined by the expression (1.33).

Different types of fuzzy relations satisfy different conditions (1.35) – (1.45). Let us consider some types of fuzzy binary relations on the finite set X .

Definition 1.25. *The fuzzy equivalence relation is the fuzzy binary relation which possesses the symmetry property (1.41), the reflexivity property (1.35), and the (max-min) transitivity property (1.44).*

This fuzzy relation is also called a similarity relation and denoted by S.

Definition 1.26. *The fuzzy dissimilarity relation is the fuzzy binary relation which possesses the symmetry property (1.41), the antireflexivity property (1.38), and the (min-max) transitivity property (1.45).*

This type of the fuzzy relations is denoted by D. Obviously, the fuzzy dissimilarity relation D is the complement of the similarity relation S on the set X . In other words, $D = \overline{S}$ and $S = \overline{D}$.

Definition 1.27. *The fuzzy tolerance relation is the fuzzy binary intransitive relation which possesses the symmetry property (1.41) and the reflexivity property (1.35).*

This type of the fuzzy relation is denoted by T . Obviously, the transitive closure $\widetilde{T}$ of some fuzzy tolerance relation T on X is the similarity relation S on X . The notions of a powerful fuzzy tolerance relation, feeble fuzzy tolerance relation and strict feeble fuzzy tolerance relations were considered in [110] and [114], as well. In particular, the powerful fuzzy tolerance relationis the fuzzy binary intransitive relation which possesses the symmetry property (1.41) and the strong reflexivity property (1.36). This type of fuzzy tolerance relationis denoted by T_3.

On the other hand, the feeble fuzzy tolerance relationis the fuzzy binary intransitive relation which possesses the symmetry property (1.41) and the weak reflexivity property (1.37). This type of the fuzzy tolerance relations is denoted by T_1. In this context the classical fuzzy tolerance relation in the sense of definition 1.27 is called the usual fuzzy tolerance relation and this type of the fuzzy tolerance relation is denoted by T_2. Note that some feeble fuzzy tolerance relation T_1 on X can be a subnormal fuzzy relation and this fact was demonstrated in [110]. The complement of some fuzzy tolerance relation T on X is the fuzzy intolerance relation I on X.

Definition 1.28. *The weak fuzzy preference relation is the fuzzy binary nonsymmetric relation which possesses the reflexivity property (1.35).*

This type of fuzzy relations is denoted by R. The weak fuzzy preference relation R is the nonsymmetric fuzzy relation in the sense that there are $x_i, x_j \in X$ such that $\mu_R(x_i, x_j) \neq \mu_R(x_j, x_i)$ which is not antisymmetric and so is not perfectly antisymmetric either.

The fuzzy likeness relation L which corresponds to the weak fuzzy preference relation R is the fuzzy relation on X with the membership function $\mu_L(x_i, x_j)$ which is defined as

$$\mu_L(x_i, x_j) = \max\left(\min(1 - \mu_R(x_i, x_j), 1 - \mu_R(x_j, x_i)), \min(\mu_R(x_i, x_j), \mu_R(x_j, x_i))\right) . \quad (1.50)$$

The fuzzy quasi-equivalence relation Q which corresponds to the weak fuzzy preference relation R is the binary fuzzy relation on X with the membership function $\mu_Q(x_i, x_j)$ which is defined by the expression

$$\mu_Q(x_i, x_j) = \min\left(\mu_R(x_i, x_j), \mu_R(x_j, x_i)\right). \quad (1.51)$$

Note that fuzzy relations L and Q were considered by Orlovsky [82], and both types of fuzzy relations are reflexive and symmetric fuzzy relations but these fuzzy relations are not equal. In other words, $L \neq Q$ for the same weak fuzzy preference relation R.

The strong fuzzy preference relation P which corresponds to the weak fuzzy preference relation R is the binary fuzzy relation on X with the membership function $\mu_P(x_i, x_j)$ which is defined as

$$\mu_P(x_i,x_j)=\begin{cases}\mu_R(x_i,x_j)-\mu_R(x_j,x_i), & if \quad \mu_R(x_i,x_j)>\mu_R(x_j,x_i)\\ 0, & if \quad \mu_R(x_i,x_j)\le\mu_R(x_j,x_i)\end{cases}. \quad (1.52)$$

Obviously, the strong fuzzy preference relation P possesses the perfect antisymmetry property (1.41) and the antireflexivity property (1.35).

The interconnection between the weak fuzzy preference relations, fuzzy likeness relations, fuzzy quasi-equivalence relations, and strong fuzzy preference relations can be explained by following example. Let $X=\{x_1,\ldots,x_4\}$ be a universe and R be a weak fuzzy preference relationon X with the membership function $\mu_R(x_i,x_j)$. The fuzzy relation R is presented in table 1.2.

Table 1.2 A weak fuzzy preference relation R

R	x_1	x_2	x_3	x_4
x_1	1.0	0.2	0.3	0.1
x_2	0.5	1.0	0.2	0.6
x_3	0.1	0.6	1.0	0.3
x_4	0.6	0.1	0.5	1.0

The fuzzy likeness relation L which corresponds to the weak fuzzy preference relation R of Table 1.2 is presented in Table 1.3.

Table 1.3 A fuzzy likeness relation L

L	x_1	x_2	x_3	x_4
x_1	1.0	0.5	0.7	0.4
x_2	0.5	1.0	0.4	0.4
x_3	0.7	0.4	1.0	0.5
x_4	0.4	0.4	0.5	1.0

Table 1.4 presents the fuzzy quasi-equivalence relation Q with the membership function $\mu_Q(x_i,x_j)$ obtained from the weak fuzzy preference relation R of Table 1.2 using the formula (1.51).

Table 1.4 A fuzzy quasi-equivalence relation Q

Q	x_1	x_2	x_3	x_4
x_1	1.0	0.2	0.1	0.1
x_2	0.2	1.0	0.2	0.1
x_3	0.1	0.2	1.0	0.3
x_4	0.1	0.1	0.3	1.0

The strong fuzzy preference relation P obtained from the weak fuzzy preference relation R of Table 1.2 using the formula (1.52) is given in Table 1.5.

Table 1.5 A strong fuzzy preference relation P

P	x_1	x_2	x_3	x_4
x_1	0.0	0.0	0.2	0.0
x_2	0.3	0.0	0.0	0.5
x_3	0.0	0.4	0.0	0.0
x_4	0.5	0.0	0.2	0.0

Other types of fuzzy relations can be obtained from the presented fuzzy relations. In particular, the fuzzy equivalence relations can be obtained by the (max-min) transitive closure of the fuzzy likeness relation or the fuzzy quasi-equivalence relation, and the corresponding fuzzy dissimilarity relations can be obtained from thesefuzzy equivalence relations using the operation of complementation (1.26). On the other hand, the fuzzy intolerance relations can be obtained from the fuzzy likeness relation and the fuzzy quasi-equivalence relation immediately as the complements of these fuzzy relations.

Let us consider the problem of decomposition of fuzzy relations which is very important from the theoretical and pratical points of view. Let R be a fuzzy relationon X with $\mu_R(x_i, x_j)$ being its membership function and R_{α_1}, R_{α_2} are its α-level sets for the corresponding values $\alpha_1, \alpha_2 \in [0,1]$. The condition

$$\alpha_2 \geq \alpha_1 \Rightarrow R_{\alpha_2} \subseteq R_{\alpha_1}, \tag{1.53}$$

is met for different α-levels of the same fuzzy relation R. So, the theorem of decomposition for the fuzzy relations was formulated by Kaufmann [60] as follows.

Theorem 1.2. *Any fuzzy relation R on X can be represented as follows:*

$$\mu_R(x_i, x_j) = \max_{\alpha_\ell}(\alpha_\ell \cdot \mu_{R_{\alpha_\ell}}(x_i, x_j)), \tag{1.54}$$

where

$$\mu_{R_{\alpha_\ell}}(x_i, x_j) = \begin{cases} 1, \textit{if } \mu_R(x_i, x_j) \geq \alpha_\ell \\ 0, \textit{if } \mu_R(x_i, x_j) < \alpha_\ell \end{cases}, \tag{1.55}$$

for all $\alpha_\ell \in [0,1]$.

The notion of a global projection of the binary fuzzy relation on the set X can be useful for the decomposition of subnormal fuzzy relations. So, the following proposition was formulated and proved in [118].

Lemma 1.1. *Let $X = \{x_1, \ldots, x_n\}$ be the finite universe and R be a fuzzy relation on X with $\mu_R(x_i, x_j)$ being its membership function. The fuzzy relation R on X can be represented using the formula (1.54) for all* $\alpha_\ell \in (0, \mathrm{Proj}(R)]$.

Proof. The formula (1.23) can be rewritten as $\mathrm{Proj}(R) = \max_{x_i, x_j} \mu_R(x_i, x_j)$, $\forall x_i, x_j \in X$. If the fuzzy relation R is the subnormal fuzzy relation, then $\mathrm{Proj}(R) < 1$. So, the values of the membership function $\mu_R(x_i, x_j)$ are absent on the interval $(\mathrm{Proj}(R), 1]$ and $\alpha_\ell \notin (\mathrm{Proj}(R), 1]$. On the other hand, if the fuzzy relation R is the normal fuzzy relation, then $\mathrm{Proj}(R) = 1$ and $\alpha_\ell \in (0, \mathrm{Proj}(R) = 1]$. □

On the other hand, the concept of the α-level fuzzy relation can be taken into account while considering the decomposition of fuzzy relations. Let $X = \{x_1, \ldots, x_n\}$ be a finite universe and R be a fuzzy relation on X with $\mu_R(x_i, x_j)$ being its membership function. An ordered sequence $0 < \alpha_0 < \ldots < \alpha_\ell < \ldots < \alpha_Z \leq 1$ of the threshold values $\alpha_\ell \in (0,1]$ must be constructed for solving the problem of decomposition of the fuzzy relations. The method of constructing the sequence $0 < \alpha_0 < \ldots < \alpha_\ell < \ldots < \alpha_Z \leq 1$ was proposed in [118].

The first value α_0 in the sequence $0 < \alpha_0 < \ldots < \alpha_\ell < \ldots < \alpha_Z \leq 1$ can be calculated as follows:

$$\alpha_0 = \min_{x_i, x_j} \mu_R(x_i, x_j), \quad \mu_R(x_i, x_j) \in (0,1], \ \forall x_i, x_j \in X, \tag{1.56}$$

and the following values can be revealed in the sequence $0 < \alpha_0 < \ldots < \alpha_\ell < \ldots < \alpha_Z \leq 1$ using the formula

$$\alpha_\ell = \min_{x_i, x_j} \mu_R(x_i, x_j), \quad \mu_R(x_i, x_j) \in (\alpha_q, \alpha_Z], \ \forall x_i, x_j \in X, \tag{1.57}$$

where $\alpha_q \in (0,1]$ is the current value of α, and the last value α_Z in the sequence $0 < \alpha_0 \leq \alpha_1 \leq \ldots \leq \alpha_\ell \leq \ldots \leq \alpha_Z \leq 1$ is determined as follows:

$$\alpha_Z = \mathrm{Proj}(R). \tag{1.58}$$

Thus, the ordered sequence $0 < \alpha_0 < \ldots < \alpha_\ell < \ldots < \alpha_Z \leq 1$ can be constructed immediately from the values of the membership function $\mu_R(x_i, x_j)$ using the formulae (1.56) – (1.58) and the corresponding fast algorithm is presented in [118].

Note that the condition (1.53) can be rewritten for the α-level fuzzy relations as follows:

$$\alpha_2 \geq \alpha_1 \Rightarrow R_{(\alpha_2)} \subseteq R_{(\alpha_1)}, \tag{1.59}$$

and the condition

$$R_{(\alpha_\ell)} \subseteq R, \tag{1.60}$$

is met for all $\alpha_\ell \in (0,1]$. The following property was formulated and proved in [117].

Lemma 1.2. *Let R be a fuzzy relation on the finite universe X with $\mu_R(x_i, x_j)$ being its membership function and $0 < \alpha_0 \leq \alpha_1 \leq \ldots \leq \alpha_\ell \leq \ldots \leq \alpha_Z \leq 1$ be an ordered sequence of the threshold values. If the value α_0 is calculated by the expression (1.56), then a condition $R_{(\alpha_0)} = R$ is met.*

Proof. The proof results immediately from the definition 1.17 and formulae (1.56), (1.59), and (1.60). □

Thus, the α -level fuzzy relations $R_{(\alpha_\ell)}$ can be constructed from the fuzzy relation R for all $\alpha_\ell \in (0,1]$. The theorem of decomposition for the fuzzy relations was formulated in [117].

Theorem 1.3. *Let* $X = \{x_1, \ldots, x_n\}$ *be a finite universe and* R *be a fuzzy relation on* X *with* $\mu_R(x_i, x_j)$ *being its membership function. The fuzzy relation* R *on* X *can be represented as follows:*

$$\mu_R(x_i, x_j) = \max_{\alpha_\ell}(\mu_{R_{(\alpha_\ell)}}(x_i, x_j)), \tag{1.61}$$

where the membership function $\mu_{R_{(\alpha_\ell)}}(x_i, x_j)$ *is defined by (1.25) for all* $\alpha_\ell \in (0,1]$.

Proof. Let $R_{(\alpha_\ell)}$ be fuzzy relations on X with their membership functions $\mu_{R_{(\alpha_\ell)}}(x_i, x_j)$ for all threshold values $\alpha_\ell \in (0,1]$. So, $\mu_R(x_i, x_j) = \mu_{R_{(\alpha_0)} \cup \ldots \cup R_{(\alpha_Z)}}(x_i, x_j) = \max_{\alpha_\ell}(\mu_{R_{(\alpha_\ell)}}(x_i, x_j))$. On the other hand, the condition $\max_{\alpha_\ell}(\mu_{R_{(\alpha_\ell)}}(x_i, x_j)) = \mu_{R_{(\alpha_0)}}(x_i, x_j)$ is met for all $\alpha_\ell \in (0,1]$ and the condition $\mu_{R_{(\alpha_0)}}(x_i, x_j) = \mu_R(x_i, x_j)$ is following directly from lemma 1.2. That is why the theorem is proved. □

Different extensions for fuzzy relations were proposed. For example, vague relations were considered by Khan, Ahmad and Biswas in [63]. Type-two fuzzy relations are considered also by different researchers. On the other hand, intuitionistic fuzzy relations were considered by Burillo and Bustince [14] and [15].

1.1.3 Fuzzy Numbers

Fuzzy intervals and fuzzy numbers can be considered as a special kind of fuzzy sets. Fuzzy numbers are useful for constructing a possibility distribution in some problems which will be considered below.

Usually, LR -type fuzzy intervals and LR -type fuzzy numbers are used to represent fuzzy data. So, the concept of an LR -type fuzzy interval and the concept of an LR -type fuzzy number should be defined in the first place. These concepts were considered, for example, by Yang and Ko in [164].

Let L or R be decreasing, shape functions from $\Re^+$ to $[0,1]$ with $L(0) = 1$ and $\forall x > 0$, $L(x) < 1$, $\forall x < 1$, $L(x) > 0$; $L(1) = 0$ or $L(x) > 0$, $\forall x$ and

$L(+\infty)=0$. Then a fuzzy set V is called an LR-type fuzzy interval $V=(\underline{m},\overline{m},a,b)_{LR}$ with $a>0$, $b>0$ if the membership function $\mu_V(x)$ of V is defined as

$$\mu_V(x)=\begin{cases} L\left(\dfrac{\underline{m}-x}{a}\right), & x\le \underline{m} \\ 1, & \underline{m}\le x\le \overline{m}, \\ R\left(\dfrac{x-\overline{m}}{b}\right), & x\ge \overline{m} \end{cases} \tag{1.62}$$

where $\underline{m}$ is called the lower mean value of V and $\overline{m}$ is called the upper mean value of V. The parameters a and b are called the left and right spreads, respectively.

For an LR-type fuzzy interval $V=(\underline{m},\overline{m},a,b)_{LR}$, if L and R are of the form

$$T(x)=\begin{cases} 1-x, & 0\le x\le 1 \\ 0, & otherwise \end{cases}, \tag{1.63}$$

then V is called a trapezoidal fuzzy interval.

The trapezoidal fuzzy interval will be denoted by $V=(\underline{m},\overline{m},a,b)_{TI}$ and its membership function is defined as follows:

$$\mu_V(x)=\begin{cases} 1-\dfrac{\underline{m}-x}{a}, & x\le \underline{m} \\ 1, & \overline{m}\le x\le \underline{m}. \\ 1-\dfrac{x-\overline{m}}{b}, & x\ge \overline{m} \end{cases} \tag{1.64}$$

Let $V=(\underline{m},\overline{m},a,b)_{LR}$ be an LR-type fuzzy interval. If the condition $\underline{m}=\overline{m}=m$ is met, then an LR-type fuzzy interval V is called an LR-type fuzzy number and its membership function is defined as

$$\mu_V(x)=\begin{cases} L\left(\dfrac{m-x}{a}\right), & x\le m \\ R\left(\dfrac{x-m}{b}\right), & x\ge m \end{cases}, \tag{1.65}$$

where m is called the mean value of V and a and b are called the left and right spreads. A fuzzy number of LR-type is denoted by $V=(m,a,b)_{LR}$.

Among the LR-type fuzzy numbers, the triangular and Gaussian fuzzy numbers are most commonly used. In particular, for an LR-type fuzzy number $V=(m,a,b)_{LR}$ if L and R are of the form

$$T(x)=\begin{cases}1-x, 0\le x\le 1\\ 0, \quad otherwise\end{cases}, \tag{1.66}$$

then V is called a triangular fuzzy number, denoted by $V=(m,a,b)_T$ and its membership function is defined as

$$\mu_V(x)=\begin{cases}1-\dfrac{m-x}{a}, x\le m\\ 1-\dfrac{x-m}{b}, x\ge m\end{cases}. \tag{1.67}$$

Let us consider the definition of the Gaussian fuzzy numbers. If $L(x)=R(x)=\exp\left(-((x-m)/\sigma)^2\right)$ for an LR-type fuzzy number $V=(m,a,b)_{LR}$, then V is called a Gaussian fuzzy number, denoted by $V=(m,\sigma)_G$. The membership function of the Gaussian fuzzy number $V=(m,\sigma)_G$ is defined as

$$\mu_V(x)=\exp\left(-\frac{(x-m)^2}{\sigma^2}\right), -\infty<x<\infty. \tag{1.68}$$

The trapezoidal fuzzy interval $V_1=(\underline{m}=12.8,\overline{m}=14.6,a=1.1,b=1.2)_{TI}$, the triangular fuzzy number $V_2=(m=24.0,a=0.9,b=1.0)_T$, and the Gaussian fuzzy number $V_3=(m=39.4,\sigma=1.2)_G$ are shown in Figure 1.2.

The triangular fuzzy numbers can be considered as a special kind of the trapezoidal fuzzy intervals. Moreover, the trapezoidal fuzzy intervals are often called the trapezoidal fuzzy numbers.

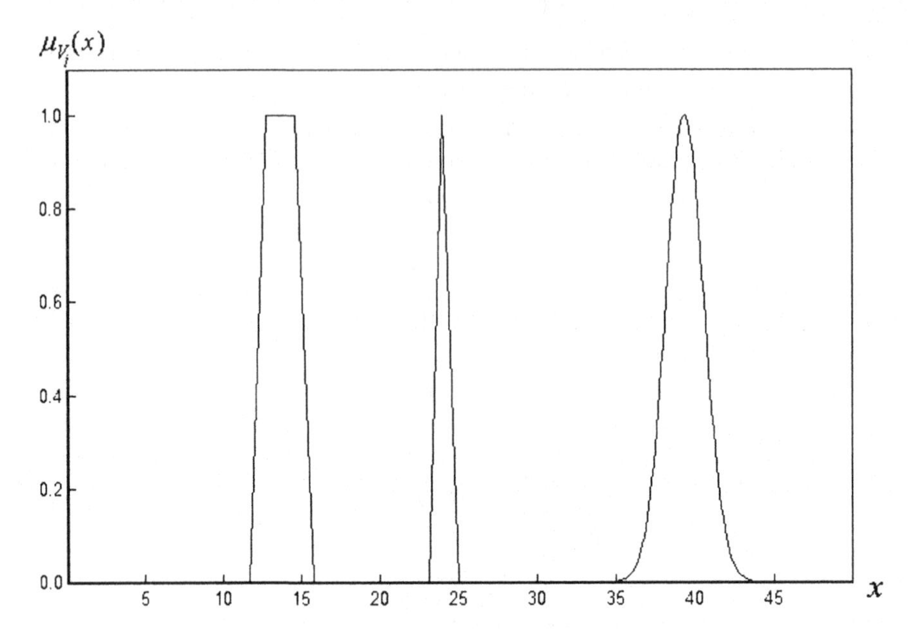

Fig. 1.2 The trapezoidal fuzzy interval and two kinds of fuzzy numbers

1.2 Basic Methods of Fuzzy Clustering

Objective function-based fuzzy clustering algorithms are considered in the first subsection which also deals with the problems of cluster validity and interpretation of clustering results. In the second subsection heuristic algorithms of fuzzy clustering are described. The third subsection of the section provides a brief description of some hierarchical fuzzy clustering procedures.

1.2.1 Optimization Methods of Fuzzy Clustering

Fuzzy clustering methods aim at discovering a suitable fuzzy partition or fuzzy coverage for a given data set. For these structures an object is not assigned to a unique cluster but has its membership degrees between zero and one to each cluster. So, the membership degrees give information about the uncertainty of the classification.

The most widespread approach in fuzzy clustering is the optimization approach. Most optimization fuzzy clustering algorithms aim at minimizing an objective function that evaluates the partition of the data into a given number of fuzzy clusters. The objective function-based fuzzy clustering algorithms can in general be divided into two types: object and relational.

The objective data clustering methods can be applied if the objects are represented as points in some multidimensional space $I^{m_1}(X)$. In other words, the data which is composed of n objects and m_1 attributes is denoted as

$\hat{X}_{n\times m_1} = [\hat{x}_i^{t_1}]$, $i = 1,\ldots,n$, $t_1 = 1,\ldots,m_1$ and the data are called sometimes the two-way data [102]. Let $X = \{x_1,\ldots,x_n\}$ is the set of objects. So, the two-way data matrix can be represented as follows:

$$\hat{X}_{n\times m_1} = \begin{pmatrix} \hat{x}_1^1 & \hat{x}_1^2 & \ldots & \hat{x}_1^{m_1} \\ \hat{x}_2^2 & \hat{x}_2^2 & \ldots & \hat{x}_2^{m_1} \\ \ldots & \ldots & \ldots & \ldots \\ \hat{x}_n^1 & \hat{x}_n^2 & \ldots & \hat{x}_n^{m_1} \end{pmatrix}. \tag{1.69}$$

Therefore, the two-way data matrix can be represented as $\hat{X} = (\hat{x}^1,\ldots,\hat{x}^{m_1})$ using n-dimensional column vectors $\hat{x}^{t_1}$, $t_1 = 1,\ldots,m_1$, composed of the elements of the t_1-th column of $\hat{X}$.

In the relational approach to fuzzy clustering, the problem of the data classification is solved by expressing a relation which quantifies either similarity or dissimilarity between pairs of objects. So, the data matrix takes the form

$$\rho_{n\times n} = \begin{pmatrix} \rho_{11} & \rho_{12} & \cdots & \rho_{1n} \\ \rho_{21} & \rho_{22} & \cdots & \rho_{2n} \\ \ldots & \ldots & \ldots & \ldots \\ \rho_{n1} & \rho_{n2} & \cdots & \rho_{nn} \end{pmatrix}, \tag{1.70}$$

where the general notation ρ_{ij} used for the designation of pair wise dissimilarity coefficients $d(x_i, x_j)$ or similarity coefficients $r(x_i, x_j)$. In general, the values ρ_{ij} are not normalized.

The traditional optimization methods of fuzzy clustering are based on the concept of a fuzzy c-partition which was introduced by Ruspini [97]. The initial set $X = \{x_1,\ldots,x_n\}$ of n objects represented by the matrix of similarity coefficients, the matrix of dissimilarity coefficients or the matrix of object attributes, should be divided into c fuzzy clusters. Namely, the grade u_{li}, $1 \le l \le c$, $1 \le i \le n$, to which an object x_i belongs to the fuzzy cluster A^l should be determined. For each object x_i, $i = 1,\ldots,n$, the grades of membership should satisfy the conditions of the fuzzy c-partition:

$$\sum_{l=1}^{c} u_{li} = 1, \ 1 \le i \le n, \ 0 \le u_{li} \le 1, \ 1 \le l \le c. \tag{1.71}$$

In other words, the family of fuzzy sets $P(X) = \{A^l \mid l = \overline{1,c}, c \le n\}$ is the fuzzy c-partition of the initial set of objects $X = \{x_1, ..., x_n\}$ if the condition (1.71) is met. A fuzzy c-partition $P(X)$ may be described using a partition matrix $P_{c \times n} = [u_{li}]$, $l = 1, \dots, c$, $i = 1, \dots, n$. The set of all fuzzy c-partitions will be denoted by Π. So, the fuzzy problem formulation in cluster analysis can be set as the optimization task $Q \to \underset{P(X) \in \Pi}{extr}$ under the constraints (1.71), where Q is a fuzzy objective function.

The optimization approach is also called a probabilistic fuzzy clustering because the condition (1.71) is met; with this constraint the membership degrees for an object may be viewed formally as the probabilities of its being a member of the corresponding clusters.

The best known optimization approach to fuzzy clustering is the method of fuzzy c-means. A fuzzy version of the well-known ISODATA-algorithm was proposed by Dunn [35], and later, Bezdek [10] generalized the fuzzy version into the fuzzy *c*-means (FCM) algorithm by introducing the weighting exponent.

The FCM-algorithm is based on an iterative optimization of the fuzzy objective function, which takes the form:

$$Q_{FCM}(P, \mathrm{T}) = \sum_{l=1}^{c} \sum_{i=1}^{n} u_{li}^{\gamma} d^2(x_i, \tau^l), \tag{1.72}$$

where u_{li}, $l = 1, \dots, c$, $i = 1, \dots, n$ is the membership degree, x_i, $i \in \{1, \dots, n\}$ is the data point, $\mathrm{T} = \{\tau^1, \dots, \tau^c\}$ is the set fuzzy clusters prototypes, $\gamma > 1$ is the weighting exponent, and $d^2(x_i, \tau^l)$ is the squared Euclidean distance between x_i and τ^l:

$$d^2(x_i, \tau^l) = \left\| x_i - \tau^l \right\|^2. \tag{1.73}$$

The purpose of the classification task is to obtain the solutions $P(X)$ and $\tau^1, \dots, \tau^c$ which minimize (1.72).

Notice that the equation (1.72) is the most popular form of the objective function in case of fuzzy clustering. Special forms of objective functions depend on the choice of a suitable distance measure. For example, the so-called alternative fuzzy c-means was proposed by Wu and Yang [156].

Thus, the FCM-algorithm is a basis of a family of fuzzy clustering algorithms. The family of objective function-based fuzzy clustering algorithms includes:

- fuzzy c-varieties algorithm (FCV): detection of linear manifolds;
- fuzzy c-lines algorithm (FCL): detection of lines;
- fuzzy c-elliptotypes algorithm (FCE): detection of hyperplanes.

On the other hand, some special cases of objective function-based fuzzy clustering algorithms can also be considered. In the first place, Gustafson and Kessel [44] designed a fuzzy clustering technique that is able to adapt to hyper-ellipsoidal forms. The Gustaffson and Kessel (GK) algorithm replaces the Euclidean distance by the Mahalanobis distance

$$d^2(x_i,\tau^l) = (\omega_l \det C_l)^{1/m_1} \cdot (x_i - \tau^l)^{\mathrm{T}} C_l^{-1}(x_i - \tau^l), \tag{1.74}$$

where the factor ω_l can be used to determine the size of the cluster A^l, $l \in \{1,\ldots,c\}$. The covariance matrices are computed by using the formula

$$C_l = \sum_{i=1}^{n} u_{li}^{\gamma}(x_i - \tau^l)(x_i - \tau^l)^{\mathrm{T}}. \tag{1.75}$$

The factor $(\omega_l \det C_l)^{1/m_1}$ in the formula (1.74) guarantees the volume for all clusters to be constant. So, the ellipsoidal clusters with approximately the same size can be obtained from the GK-algorithm. There isan axis-parallel variant of the GK-algorithm, the AGK-algorithm, which can also be used to detect lines [49].

In the second place, Gath and Geva [39] proposed the extension of the GK-algorithm, the GG-algorithm, which is not based on an objective function optimizer. Their GG-algorithm is a heuristic method derived from the fuzzification of the maximum likelihood estimator. The main idea is to assume that the objects are part of a m_1-dimensional normal distribution. So, for the GG-algorithm, the distance measure is of the following form:

$$d^2(x_i,\tau^l) = \frac{1}{p_l}\sqrt{\det(D_l)} \cdot \exp^{\left(\frac{1}{2}(x_i-\tau^l)^{\mathrm{T}} D_l^{-1}(x_i-\tau^l)\right)}, \tag{1.76}$$

where the parameter p_l denotes the a priori probability for an object to belong to the l-th normal distribution and the parameter can be estimated by using

$$p_l = \frac{\sum_{i=1}^{n} u_{li}^{\gamma}}{\sum_{a=1}^{c}\sum_{i=1}^{n} u_{ai}^{\gamma}}. \tag{1.77}$$

On the other hand, the covariance matrix D_l of the l-th normal distribution can be estimated by using

$$D_l = \frac{\sum_{i=1}^{n} u_{li}^{\gamma}(x_i - \tau^l)(x_i - \tau^l)^{\mathrm{T}}}{\sum_{i=1}^{n} u_{li}^{\gamma}}. \tag{1.78}$$

Thus, the ellipsoidal clusters with a varying size can be obtained from the GG-algorithm. There is also an axis-parallel version of this algorithm, the AGG-algorithm, which can be used to detect lines.

Thirdly, an approach to fuzzy membership assignments by regularizing the objective function of crisp clustering was proposed by Li, Miyamoto, and Mukaidono [79] and [71]. The fuzzification technique is called "regularization by entropy" and the following objective function is used:

$$Q_{MEC}(P,\mathrm{T}) = \sum_{l=1}^{c}\sum_{i=1}^{n} u_{li} d^2(x_i,\tau^l) + \lambda \sum_{l=1}^{c}\sum_{i=1}^{n} u_{li} \log u_{li}, \tag{1.79}$$

where the entropy term is added for the fuzzification instead of the weighting exponent in the standard FCM-algorithm and $d^2(x_i,\tau^l)$ is the squared Euclidean distance (1.73).

The MEC-algorithm name was used by Wang, Chung, Deng, Hu and Wu [152] for the model and algorithm because the entropy-based fuzzy clustering method is called also the maximum entropy clustering method. So, the coefficient λ works like the fuzzifier γ of the FCM-algorithm. The larger λ the fuzzier the membership assignments are. The MEC-algorithm is the basic algorithm for the entropy-based fuzzy clustering method which wa developed by Miyamoto, Ichihashi and Honda [80]. It is should be noted that the entropy-based fuzzy clustering method is called sometimes as the Gaussian-clustering method. That is why the MEC-algorithm is called the GCM-algorithm in [71].

In contrast to the methods that are elaborated for solid cluster detection, the so-called shell-clustering algorithms are tailored for clusters in the form of boundaries of circles or ellipses. The family of shell-clustering algorithms includes:

- fuzzy c -shells algorithm (FCS): detection of circles;
- fuzzy c -ellipsoidal shells algorithm (FCES): detection ofellipsoids;
- fuzzy c -quadric shells algorithm (FCQS): detection of ellipsoids.

In the algorithms, each prototype $\tau^l = (\tau^l, R_l)$ consists of the cluster centre τ^l and the radius R_l. For example, the distance function for the FCS-algorithm is

$$d^2(x_i,\tau^l) = \left(\left\|x_i - \tau^l\right\| - R_l\right)^2, \tag{1.80}$$

where the radius is calculated using the formula

$$R_l = \frac{\sum_{i=1}^{n} u_{li}^{\gamma} \left\| x_i - \tau^l \right\|}{\sum_{i=1}^{n} u_{li}^{\gamma}}, \tag{1.81}$$

for all fuzzy clusters A^l, $l = 1, \ldots, c$. However, the distance function (1.80) leads to a set of coupled non-linear equations for the τ^l and R_l that cannot be solved in an analytic way.

On the other hand, the FCRS-algorithm for the detection of clusters in the form of boundaries of rectangles was developed by Höppner [48]. All objective function-based fuzzy clustering algorithms were proposed by different authors and the clustering procedures are described in detail by Höppner, Klawonn, Kruse and Runkler [49].

The relational clustering algorithms generate a fuzzy c-partition from relational data based on the minimization of an objective function. These objective function assumes ρ_{ij} to be a pairwise dissimilarity relation between objects in the set X. The approach was outlined by Ruspini [98]. The most popular examples of the fuzzy relational clustering are the RFCM-algorithm which was described by Hathaway, Davenport, and Bezdek [45], Roubens' FNM-algorithm [95], Windham's AP-algorithm [155], and the ARCA-algorithm which was proposed by Corsini, Lazzerini, and Marcelloni [25]. Let us consider the objective functions of these algorithms.

First, the relational dual of the FCM-algorithm was proposed in [45]. So, the RFCM-algorithm is based on the objective function which takes the form

$$Q_{RFCM}(P) = \sum_{l=1}^{c} \frac{\sum_{j=1}^{n}\sum_{i=1}^{n} u_{li}^{\gamma} u_{lj}^{\gamma} d(x_i, x_j)}{2\sum_{k=1}^{n} u_{lk}^{\gamma}}, \tag{1.82}$$

where the conditions $d(x_i, x_j) \geq 0$, $d(x_i, x_i) = 0$, $d(x_i, x_j) = d(x_j, x_i)$ and $d(x_j, x_k) = \left\| x_j - x_k \right\|^2$ are met for all $x_i, x_j, x_k \in X$. The RFCM-algorithm assumes that the relation matrix $\rho_{n \times n} = [d(x_i, x_j)]$ is obtained from the distances between pairs of two-way data. Notice that $\rho_{n \times n}$ is not necessarily a fuzzy dissimilarity relation and a relation $\rho_{n \times n}$ must satisfy the Euclidean relation requirements. In particular, a relation $\rho_{n \times n}$ is Euclidean if there exists the data set

$X=\{x_1,\ldots,x_n\}$ in the space $I^{m_1}(X)$ where $m_1=n-1$ such that $\rho_{n\times n}=\left[d(x_j,x_k)=\|x_j-x_k\|^2\right]$. Otherwise, $\rho_{n\times n}$ is said to be non-Euclidean. The corresponding two-way data set $X=\{x_1,\ldots,x_n\}$ is called a realization of the Euclidean relation $\rho_{n\times n}$. An extension of the RFCM-algorithm for some non-Euclidean relation $\rho_{n\times n}$ was proposed by Hathaway and Bezdek [47] and the extension is called the NERFCM-algorithm. Moreover, when $\gamma=2$, the equation (1.82) is the objective function of the well-known FANNY-algorithm which was proposed by Kaufmann and Rousseeuw [61].

Second, the fuzzy non-metric model was proposed by Roubens [95] and the objective function of that model can be written as follows:

$$Q_{FNM}(P)=\sum_{l=1}^{c}\sum_{i=1}^{n}\sum_{j=1}^{n}u_{li}^2u_{lj}^2d(x_i,x_j). \tag{1.83}$$

If the assumption

$$d(x_i,\tau^l)=\sum_{j=1}^{n}u_{lj}^2d(x_i,x_j), \tag{1.84}$$

is met, then the objective function in (1.83) can be rewritten as

$$Q_{FNM}(P)=\sum_{l=1}^{c}\sum_{i=1}^{n}u_{li}^2d(x_i,\tau^l), \tag{1.85}$$

which appears the same as the objective function of the FCM-algorithm in (1.72) for $\gamma=2$. So, the FNM-algorithm converges to a local minimum of the objective function in (1.83).

Third, the assignment-prototype model was presented by Windham [155]. The corresponding AP-algorithm is based on the fact that the minimum of the objective function

$$Q_{AP}(P,K)=\sum_{l=1}^{c}\sum_{i=1}^{n}\sum_{j=1}^{n}u_{li}^2v_{lj}^2d(x_i,x_j), \tag{1.86}$$

occurs at an assignment matrix, $P_{c\times n}=[u_{li}]$, and a prototype weights matrix, $K_{c\times n}=[v_{li}]$, under the constraints (1.71) and

$$\sum_{j=1}^{n}v_{lj}=1,\ \sum_{l=1}^{c}v_{lj}>0. \tag{1.87}$$

The AP-algorithm is an iterative procedure for minimizing the objective function (1.86) which combines the dissimilarity coefficients with assignment measures and prototype weights. So, the AP-algorithm is designed to find matrices $P_{c\times n} = [u_{li}]$ and $K_{c\times n} = [v_{li}]$. The clustering procedure attempts to group objects so that within a class the dissimilarities are low.

Fourth, the FCM-like relational clustering procedure was presented by Corsini, Lazzerini, and Marcelloni [25] and the algorithm was called any relation clustering algorithm, the ARCA-algorithm, in which each object is represented by the vector of its relation strengths with the other objects in the data set $X = \{x_1,..., x_n\}$, and a prototype is the point whose relationship with all the objects in the data set is representative of the mutual relationships of a group of similar objects.

So, the ARCA-algorithm is based on the criterion (1.72) under the constraints (1.71) where

$$d(x_i, \tau^l) = \sqrt{\sum_{j=1}^{n} \left(d(x_i, x_j) - d(x_j, \tau^l)\right)^2}\ , \tag{1.88}$$

and $d(x_i, x_j)$ is the dissimilarity relation between the pair of objects x_i and x_j, and $d(x_j, \tau^l)$ is the relation between the prototype τ^l, $l \in \{1,\ldots,c\}$ and the object x_j, $j \in \{1,\ldots,n\}$.

For all fuzzy clustering approaches described above the number of clusters has to be specified in advance. However, the most important problem of fuzzy clustering is neither the choice of the numerical procedure nor the distance to use but concerns the number c of fuzzy clusters to look for. Really, lacking a priori knowledge of the data structure, there is no reason to choose a particular value of c and one must find a way to measure the acceptance with which cluster structure has been identified by a clustering procedure. This is the so-called cluster validity problem.

The classical approach to cluster validity for fuzzy clustering is based on directly evaluating the fuzzy c-partition. Many measures of cluster validity can be used for the purpose. Many authors have proposed several measures of cluster validity associated with the fuzzy c-partitions. The cluster validity problem can be illustrated by the method of fuzzy c-means.

In the first place,the partition coefficient and the partition entropy were proposed by Bezdek [10]. The partition coefficient rates the crispness of a classification and the partition coefficient is defined by the following expression:

$$V_{pc}(P) = \frac{1}{n}\sum_{l=1}^{c}\sum_{i=1}^{n} u_{li}^2\ . \tag{1.89}$$

The crisper the membership values the better the classification. So, the number of clusters that maximizes the partition coefficient (1.89) is taken as the optimal number c of fuzzy clusters.

The partition entropy is similar to the partition coefficient and an optimal classification provides a minimal value for the validity measure. The partition entropy is defined as follows:

$$V_{pe}(P) = -\frac{1}{n}\sum_{l=1}^{c}\sum_{i=1}^{n} u_{li} \cdot \ln u_{li} \,. \tag{1.90}$$

Thus, the partition coefficient and the partition entropy are bounded in such a way that conditions $1/c \leq V_{pc}(P) \leq 1$ and $0 \leq V_{pe}(P) \leq \ln c$ are met.

In the second place, Xie and Beni proposed in [158] a well-known validity index which measures the overall average compactness against the separation of the fuzzy c-partition. So, the compactness and separation index is defined in [158] as follows:

$$V_{cs}(P;\Upsilon) = \frac{\sum_{l=1}^{c}\sum_{i=1}^{n} u_{li}^{\gamma} \left\| x_i - \tau^l \right\|^2}{n \cdot \min_{a \neq l} \left\| \tau^a - \tau^l \right\|^2} \,. \tag{1.91}$$

The minimum of the compactness and separation index (1.91) as a function of the number of fuzzy clusters c is sought to obtain a well-defined fuzzy c-partition $P(X)$. The compactness and separation index (1.91) is the most popular cluster validity criterion for the FCM-algorithm. Moreover, several generalizations of the compactness and separation index (1.91) were proposed and studied in [11].

The cluster validity indexes for other fuzzy clustering algorithms were proposed by different researchers and a very good overview of the indexes is given, for example, in [13]. In particular, together with the GG-algorithm, Gath and Geva proposed in [39] three validity measures. Firstly, the fuzzy hypervolume was defined by the equation

$$V_{fhv}(P;\overline{\Upsilon}) = \sum_{l=1}^{c} \sqrt{\det(D_l)} \,, \tag{1.92}$$

where the covariance matrix D_l can be calculated using the formula (1.78). To calculate the covariance matrix, only the data vectors, cluster centers and membership degrees are used. The minimum of the fuzzy hypervolume (1.92) denotes the compact fuzzy clusters.

Secondly, the average partition density can be described as the average number of data in the center of cluster weighted by the volume of cluster:

$$V_{apd}(P;\mathrm{T}) = \frac{1}{c} \cdot \sum_{l=1}^{c} \frac{\sum_{x_i \in \omega_l} u_{li}}{\sqrt{\det(D_l)}}, \tag{1.93}$$

where $\omega_l = \{x_i \in I^{m_1}(X) \mid (x_i - \tau^l)^{\mathrm{T}} D_l^{-1}(x_i - \tau^l) < 1\}$, $l = 1,\ldots,c$ is the area centered at the prototype τ^l, $l \in \{1,\ldots,c\}$. The index (1.93) should be maximized.

Thirdly, the partition density can be calculated according to the formula

$$V_{pd}(P;\mathrm{T}) = \frac{\sum_{l=1}^{c} \sum_{x_i \in \omega_l} u_{li}}{\sum_{l=1}^{c} \sqrt{\det(D_l)}}, \tag{1.94}$$

where $\omega_l = \{x_i \in I^{m_1}(X) \mid (x_i - \tau^l)^{\mathrm{T}} D_l^{-1}(x_i - \tau^l) < 1\}$, $l = 1,\ldots,c$. The partition density is equivalent to the average partition density (1.93) with the exception that the densities are summed up and the average is not evaluated. That is why the index (1.94) should be maximized too.

When the initial data are relational, the direct indices are not applicable to the problem of cluster validity. The indirect indices such as the partition coefficient (1.89) and the partition entropy (1.90) can be used for validation in these cases. On the other hand, the indirect indices can be modified for the relational clustering algorithms in each concrete case. For example, the partition coefficient (1.89) was modified by Roubens [95] as follows:

$$V_{mpc}(P) = \frac{c\left[\sum_{l=1}^{c}\sum_{i=1}^{n} u_{li}^2\right] - n}{n(c-1)}. \tag{1.95}$$

The modified partition coefficient (1.95) should be maximized. Some other validity measures for the FNM-algorithm were proposed by Libert and Roubens [72] and [73].

Notice that the compactness and separation index (1.91) is appropriate for the ARCA-algorithm, because the ARCA-algorithm, though being a relational clustering algorithm, generates prototypes. In general, the problem of cluster validity of the relational data is considered in detail by Sledge, Havens, Bezdek and Keller [103].

The basic idea of unsupervised fuzzy clustering is to define an upper bound for the number of clusters $c_{\max}$ and carry out the clustering for each number of fuzzy clusters $c \in \{2,\ldots,c_{\max}\}$. So, the validity criteria are used sequentially as follows:

1. Compute different fuzzy c -partitions for $c = 2, \dots, c_{\max}$;
2. Compute the value of a validity criterion;
3. Seek for the extreme value of validity criterion and set the optimal number of clusters to its correspondent c value.

On the other hand, some fuzzy clustering algorithms were proposed that do not require the pre-definition of the number of clusters. The CA-algorithm which was proposed by Frigui and Krishnapuram [37] and the E-FCM-algorithm which was developed by Kaymak and Setnes [62] are good examples of such clustering procedures. The basic idea of such algorithms can be outlined as follows: very small clusters are deleted and neighboring clusters are joined together; the number of fuzzy clusters is reduced and the process is repeated until no more fuzzy clusters exist which can be deleted or joined together.

Finally, a problem of the clustering results interpretation must be considered in brief. For the purpose, different approaches can be used.

Firstly, the clustering results can be converted into the matrix of a crisp partition. A defuzzification by the maximum memberships rule can be described by the formula:

$$P_{(l)}^{H} = \mathrm{e}_i \Leftrightarrow u_{li} > u_{lj}\,,\, j = 1, \dots, c\,,\ j \neq i\,. \tag{1.96}$$

If the rule (1.96) is applied to the matrix $P_{c \times n} = [u_{li}], l = 1, \dots, c$, $i = 1, \dots, n$ of the fuzzy c-partition $P(X)$, then the result of some fuzzy clustering algorithm application to the data is the matrix of a crisp partition.

Secondly, a method for a representation of the structure of fuzzy clusters was proposed by Pedrycz [89]. The method can be described in the following way. Let $X = \{x_1, \dots, x_n\}$ is the data set and the fuzzy c -partition $P(X)$ obtained from some fuzzy clustering algorithm is described with the aid of the partition matrix $P_{c \times n} = [u_{li}]$, $l = 1, \dots, c$, $i = 1, \dots, n$. The object $x_i \in X$ can be an element of the core of the l -th cluster if at least u_{li} exceeds a given threshold which will be denoted as φ. Furthermore, for each data point its structural relationship to all existing clusters should be taken into account, that is, consider $x_i \in X$ again and take also into account all membership values u_{li} of x_i to the other fuzzy clusters. If c fuzzy clusters are considered, then any object having a membership value to each of the clusters equal to $1/c$ does not contribute to a significant clustering structure. The index

$$\xi'(x_i) = -\prod_{l=1}^{c} u_{li} + \frac{1}{c^c} \tag{1.97}$$

was proposed in [89] as a measure for the structural property in x_i. Multiplying the index (1.94) by c^c, which is a constant, the normalized index

$$\xi(x_i) = 1 - c^c \prod_{l=1}^{c} u_{li} \tag{1.98}$$

is obtained, $\xi : X \to [0,1]$. From the equation (1.98) and the properties of the fuzzy c-partition matrix $P_{c\times n} = [u_{li}]$, it follows that if belongs to one of the fuzzy clusters with the membership degree equal to 1, then the normalized index (1.98) attains its maximum.

The structural significance of the core of the fuzzy cluster can be specified by the threshold value ϕ such that the object x_i can be an element of the core of the l-th fuzzy cluster if the value u_{li} exceeds the value of the threshold ϕ. Combining the foregoing, the definition of an (ϕ, φ)-core of fuzzy cluster can be formulated as follows [89].

Definition 1.29. *A subset* $X^l_{(\phi,\varphi)} \subset X$ *is said to be an* (ϕ, φ)*-core of the fuzzy cluster* $A^l \in P(X)$, $l = 1,\ldots,c$, *with* $\phi, \varphi \in [0,1]$ *if it contains the following set of objects*

$$\left\{ x_i \in X \,\middle|\, 1 - c^c \prod_{l=1}^{c} u_{li} \geq \phi,\, u_{li} \geq \varphi \right\}, \tag{1.99}$$

that the condition

$$X^l_{(\phi\varphi)} = \{x_i \in X \,|\, \xi(x_i) \geq \phi, u_{li} \geq \varphi\} \tag{1.100}$$

is met.

So, the set of (ϕ, φ)-cores $X^l_{(\phi\varphi)}$, $l = 1,\ldots,c$ and a residual data set X^R containing all remaining elements of the data set X can be obtained. Hence, the condition

$$X = \bigcup_{l=1}^{c} X^l_{(\phi\varphi)} \cup X^R, \tag{1.101}$$

is met, where the first part of the expression (1.101) represents the significant structure, and the second part corresponds to the insignificant structure of the data set X.

Thus, the cores of the fuzzy clusters controlled independently by the thresholds φ and ϕ generate a hierarchical structure of analytical representations of (ϕ, φ)-cores. The notion of the core structure of the fuzzy cluster seems as an effective tool for transparent interpretation of fuzzy clustering results.

Thirdly, the notion of α-cores of fuzzy clusters was outlined in [128]. Let $X = \{x_1, \ldots, x_n\}$ is the data set and fuzzy clusters A^l, $l = 1, \ldots, c$ are elements of the fuzzy c-partition $P(X)$ obtained from some fuzzy clustering algorithm. The fuzzy c-partition is represented by the matrix $P_{c \times n} = [u_{li}]$, $l = 1, \ldots, c$, $i = 1, \ldots, n$. This notion involves finding a threshold value $\alpha \in (0,1]$ such that the condition

$$\sum_{l=1}^{c} card\left(Supp(A^l_{(\alpha)})\right) \geq card(X), \tag{1.102}$$

is met, where an α-level fuzzy set $A^l_{(\alpha)}$ of a fuzzy set $A^l \in P(X)$, $l \in \{1, \ldots, c\}$ is the α-core of the corresponding fuzzy cluster. The membership function of α-cores of the fuzzy clusters is defined by the following expression:

$$u^{\alpha}_{li} = \begin{cases} u_{li}, \ if \ x_i \in A^l_{\alpha} \\ 0, \ \ if \ x_i \notin A^l_{\alpha} \end{cases}, \tag{1.103}$$

where the α-level A^l_{α} of the α-level fuzzy set $A^l_{(\alpha)}$, $l \in \{1, \ldots, c\}$ is the support of the corresponding α-core for a value $\alpha \in (0,1]$, $A^l_{\alpha} = Supp\left(A^l_{(\alpha)}\right)$.

The threshold value $\alpha \in (0,1]$ must be selected so that each object belongs to at least one α-core of a fuzzy cluster, and can be calculated according to the formula

$$\hat{\alpha} = \min_i \max_l u_{li} \,. \tag{1.104}$$

So, the following proposition was formulated and proved in [128].

Theorem 1.4. *Let $X = \{x_1, \ldots, x_n\}$ be the data set and $P(X) = \{A^1, \ldots, A^1\}$ be a fuzzy c-partition obtained from some clustering procedure. The coverage of the data set X can be formed by the supports $\{A^1_{\alpha}, \ldots, A^c_{\alpha}\}$ of α-cores $\{A^1_{(\alpha)}, \ldots, A^c_{(\alpha)}\}$ of fuzzy clusters if and only if the condition $\alpha \leq \hat{\alpha}$ is met, where the threshold value $\hat{\alpha}$ is calculated in accordance with the formula (1.104).*

Proof. If for some $\alpha \in (0,1]$ the family of sets $C = \{A_{\alpha}^{1},\ldots,A_{\alpha}^{c}\}$, which are α-cores of fuzzy clusters $A^{1},\ldots,A^{c}$, forms the coverage of $X = \{x_1,\ldots,x_n\}$, then each object $x_i \in X$ is an element of at least one subset $A_{\alpha}^{l} \in C$. If each object $x_i \in X$ is an element of only one subset $A_{\alpha}^{l} \in C$, then the condition $\bigcap_{l=1}^{c} A_{\alpha}^{l} = \varnothing$ is met. If each object $x_i \in X$ is an element of more than one subset $A_{\alpha}^{l} \in C$, then the condition $\bigcap_{l=1}^{c} A_{\alpha}^{l} \neq \varnothing$ is met. A set $\bigcup_{l=1}^{c} A_{\alpha}^{l}$ is the smallest set containing all sets $\{A_{\alpha}^{1},\ldots,A_{\alpha}^{c}\}$ and the condition $\bigcap_{l=1}^{c} A_{\alpha}^{l} \subseteq \bigcup_{l=1}^{c} A_{\alpha}^{l}$ is met. Hence, the condition $x_i \in \bigcup_{l=1}^{c} A_{\alpha}^{l}$, $\forall i \in \{1,\ldots,n\}$ is met.

On the other hand, since the set X is finite, then there exists at least one element $x_i \in X$ for which the condition

$$\hat{\alpha} = \min_{i} \mu_{A}(x_i), \tag{1.105}$$

is met, where A is a fuzzy set which is obtained as the fuzzy union of fuzzy sets $A^{1},\ldots,A^{c}$ in the sense of the maximum operation (1.17). Thus, the condition $\mu_{A_{(\tilde{\alpha})}}(x_i) = 0$ will be met for some value $\tilde{\alpha} > \hat{\alpha}$. Hence, the condition $x_i \notin Supp(A_{(\tilde{\alpha})})$ will be met and thecondition $x_i \notin \bigcup_{l=1}^{c} A_{\tilde{\alpha}}^{l}$ will also be satisfied. That is why the theorem is correct. □

Some propositions are corollaries of this theorem and these corollaries were formulated also in [128]. Let us consider two most important corollaries. Proofs of these corollaries are obvious and are omitted here.

Corollary 1.1. *If the equality is met in the expression (1.102), then the supports* $\{A_{\alpha}^{1},\ldots,A_{\alpha}^{c}\}$ *of the* α*-cores* $\{A_{(\alpha)}^{1},\ldots,A_{(\alpha)}^{c}\}$ *of the fuzzy clusters form a partition of a set of objects into disjoint sets.*

Corollary 1.2. *If the condition* $\alpha = \hat{\alpha}$ *is met, where the value* $\hat{\alpha}$ *is calculated using the formula (1.104), then the crisp coverage which is formed by the supports* $\{A_{\alpha}^{1},\ldots,A_{\alpha}^{c}\}$ *of the* α*-cores* $\{A_{(\alpha)}^{1},\ldots,A_{(\alpha)}^{c}\}$ *of fuzzy clusters of the fuzzy* c*-partition* $P(X) = \{A^{1},\ldots,A^{c}\}$ *is minimal.*

So, the notion of the α-cores of fuzzy clusters allows to include each object of $X = \{x_1, \ldots, x_n\}$ in the smallest number $\tilde{c}$, $1 \le \tilde{c} \le c$ of fuzzy clusters of the fuzzy c-partition $P(X) = \{A^1, \ldots, A^c\}$, which is the result of classification. On the other hand, this notion allows to save values of the membership function which can be interpreted as the degree possession of an object $x_i \in X$ of properties of class which is associated with the corresponding fuzzy cluster A^l, $l \in \{1, \ldots, c\}$. It is should be noted that the approach is simple and more useful than oftwo previous methods of the interpretationof fuzzy clustering results because the membership values of objects are saved.

The notion of α-cores of fuzzy clusters can be explained by the following simple example [128]. Let $\{V_1, \ldots, V_{30}\}$ be a set of triangular fuzzy numbers which are shown in Figure 1.3.

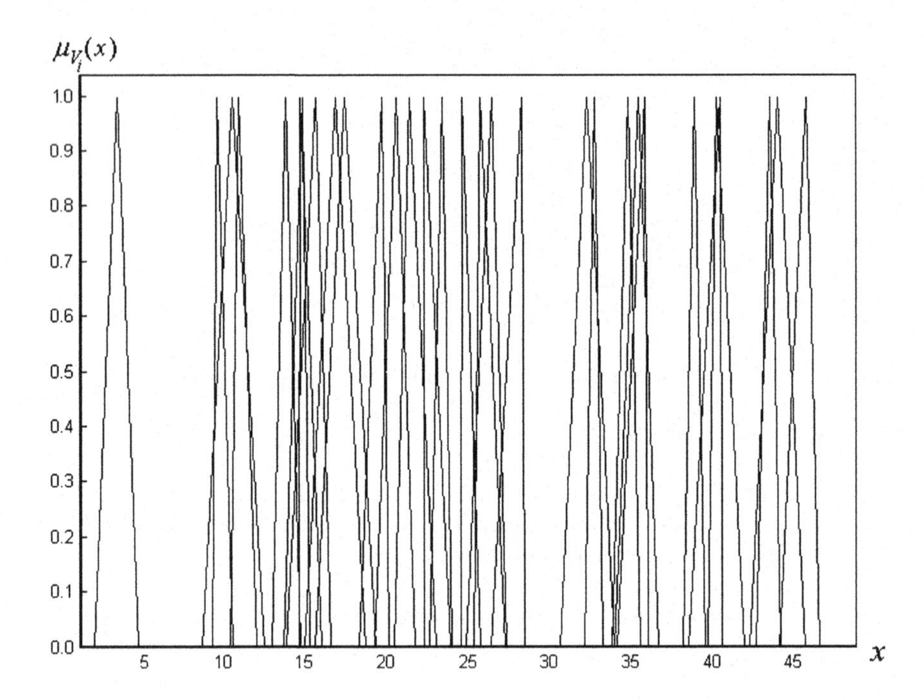

Fig. 1.3 The graph of thirty triangular fuzzy numbers

The problem of fuzzy data clustering will be considered later in detail, but it seems like an appropriate place to demonstrate an example of the notion of α-cores of fuzzy clusters. The fuzzy data can be processed by the method of fuzzy c-numbers which was proposed by Yang and Ko [164]. In general, the objective function of their FCN-algorithm takes the form (1.72) which can be rewritten here as follows:

$$Q_{FCN}(P,\mathrm{T}) = \sum_{l=1}^{c}\sum_{i=1}^{n} u_{li}^{\gamma} d^2(V_i,\tau^l), \tag{1.106}$$

where V_i, $i \in \{1,\ldots,n\}$ is the object which is represented by a fuzzy number, $\mathrm{T} = \{\tau^1,\ldots,\tau^c\}$ is the set of prototypes of fuzzy clusters which are also fuzzy numbers, $\gamma > 1$ is the weighting exponent, and $d^2(V_i,\tau^l)$ is a distance between a fuzzy number V_i and a prototype τ^l.

Let $\mathcal{F}_{(T)FN}(\Re)$ be a space of all trapezoidal fuzzy intervals and $\{V_1,\ldots,V_n\}$ be a set of n triangular fuzzy numbers in $\mathcal{F}_{(T)FN}(\Re)$. The distance $d^2_{(T)FN}(V_i,V_j)$ for any two triangular fuzzy numbers $V_i = (m_i,a_i,b_i)_T$ and $V_j = (m_j,a_j,b_j)_T$ in the space $\mathcal{F}_{(T)FN}(\Re)$ of all triangular fuzzy numbers, whose membership function are defined by the expression (1.67), was defined in [164] as follows:

$$d^2_{(T)FN}(V_i,V_j) = (m_i - m_j)^2 + \left((m_i - m_j) - \frac{1}{2}(a_i - a_j)\right)^2 + \left((m_i - m_j) + \frac{1}{2}(b_i - b_j)\right)^2. \tag{1.107}$$

So, the replacement of $d^2(V_i,\tau^l)$ with $d^2_{(T)FN}(V_i,\tau^l)$ should be made in the objective function (1.106) for clustering the data set $\{V_1,\ldots,V_n\}$ of triangular fuzzy numbers, where prototypes τ^l, $l = 1,\ldots,c$ are also triangular fuzzy numbers.

The data set was processed by the FCN-algorithm with the number of fuzzy clusters $c = 3$ and the weighting exponent value $\gamma = 2$. So, the matrix $P_{3\times 30} = [u_{li}]$, $l = 1,\ldots,3$, $i = 1,\ldots,30$ of fuzzy c-partition $P(X) = \{A^1, A^2, A^3\}$ was obtained and the membership functions of three fuzzy clusters are presented in Figure 1.4. The membership values of the first class are represented by ○, the membership values of the second class are represented by ■ and the membership values of the third class are represented by ▲.

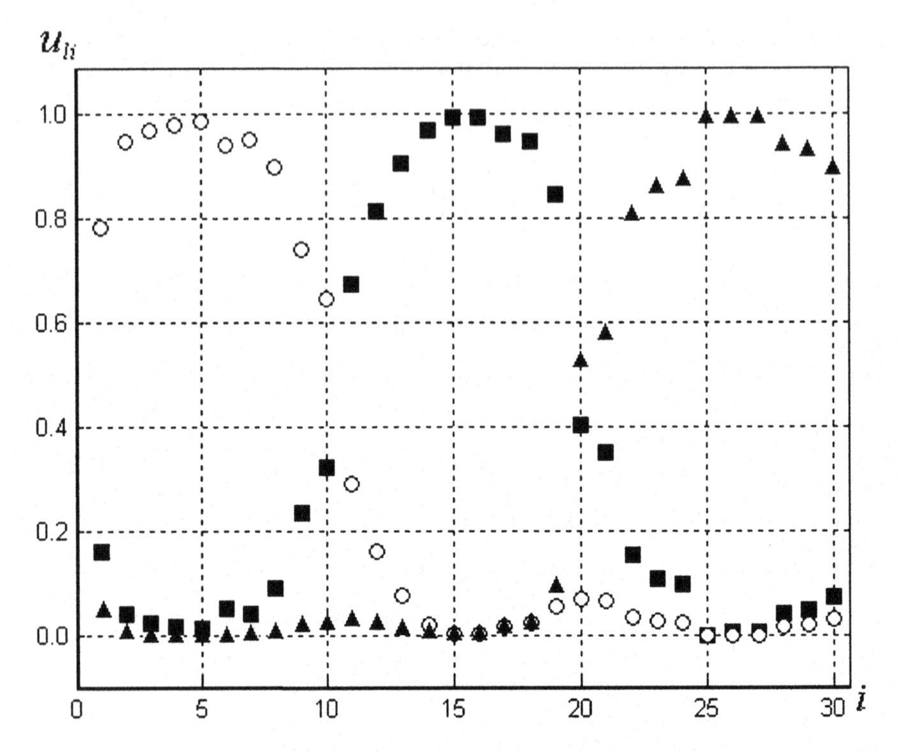

Fig. 1.4 Membership functions of three fuzzy clusters obtained from the FCN-algorithm

A set $\mathrm{T} = \{\tau^1, \ldots, \tau^3\}$ of fuzzy clusters prototypes was also obtained from the FCN-algorithm. The triangular fuzzy numbers $\tau^1 = (m^1 = 12.773, a^1 = 1.115, b^1 = 1.185)_T$, $\tau^2 = (m^2 = 24.087, a^2 = 0.914, b^2 = 1.05)_T$, and $\tau^3 = (m^3 = 39.521, a^3 = 1.283, b^3 = 0.87)_T$ are prototypes of corresponding fuzzy clusters $A^l \in P(X)$, $l = 1, \ldots, 3$.

Notice that the number of fuzzy clusters $c = 3$ was predefined in the experiment [164]. However, the result of the numerical experiment seems to be satisfactory.

The formula (1.104) was applied to the matrix $P_{3\times 30} = [u_{li}]$ of the obtained fuzzy c -partition and the threshold value $\alpha = 0.5255$ was obtained. It is should be noted that the threshold value $\alpha = 0.5255$ is equal to the membership degree of the object V_{20} to the fuzzy cluster $A^3 \in P(X)$.

The membership values of the α-cores of fuzzy clusters for the obtained threshold value α were calculated according to the formula (1.103). These membership values are shown in Figure 1.5.

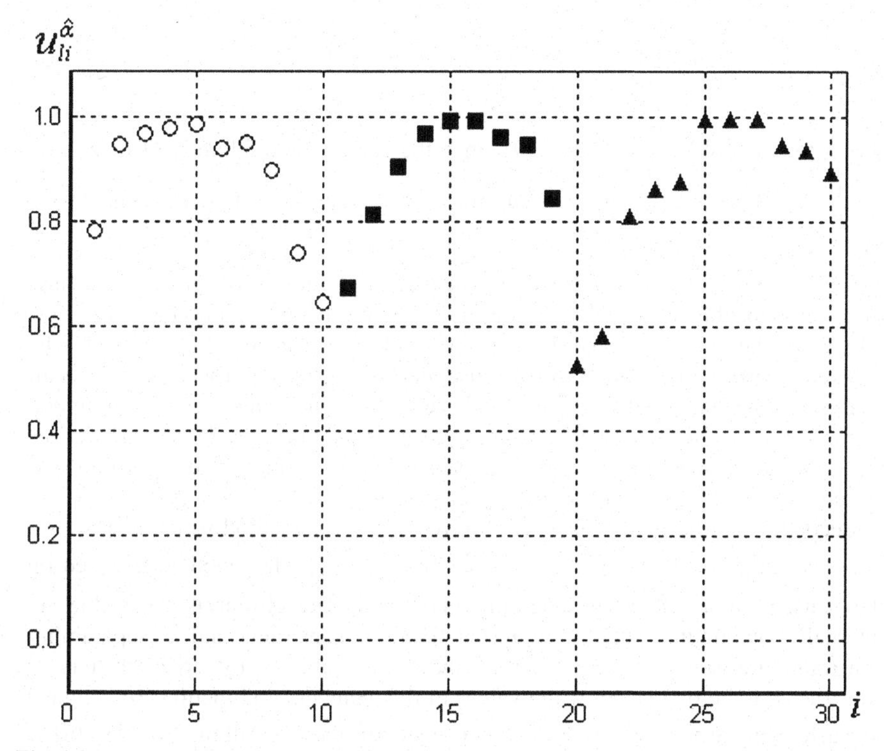

Fig. 1.5 Membership functions of α-cores of three fuzzy clusters

Obviously, the supports of the α-cores of fuzzy clusters form a crisp partition of a set of objects $\{V_1, \ldots, V_{30}\}$ into three disjoint sets.

A detailed overview on the above and other fuzzy clustering methods is given, for example, in [11], [49], and [101].

1.2.2 Heuristic Algorithms of Fuzzy Clustering

Heuristic algorithms of fuzzy clustering display as a rule a high level of essential and functional clarity and a low level of complexity. Some heuristic clustering algorithms are based on a specific definition of a cluste and the aim of those algorithms is cluster detection with respect to a given definition. As Mandel [76] has noted, such algorithms are called algorithms of direct classification or direct clustering algorithms. The direct heuristic algorithms of fuzzy clustering are simple and very effective and efficient in many cases.

The concept of a fuzzy coverage is often usedin the heuristic fuzzy clustering procedures. Let $X = \{x_1, \ldots, x_n\}$ be a set of n objects represented by the matrix of similarity coefficients, the matrix of dissimilarity coefficients or the matrix of vales of the object attributes. The set X should be divided into c fuzzy clusters. If the condition

$$\sum_{l=1}^{c} u_{li} \geq 1, \ 1 \leq i \leq n, \ 0 \leq u_{li} \leq 1, \ 1 \leq l \leq c \tag{1.108}$$

is met for each object x_i, $1 \leq i \leq n$, then the corresponding family of fuzzy sets $C(X) = \{A^l \mid l = \overline{1,c}, c \leq n\}$ is the fuzzy coverage of the initial set of objects $X = \{x_1, \ldots, x_n\}$.

Let us consider some well-known heuristic algorithms of fuzzy clustering. First, an algorithm which partitions a given sample from a multimodal fuzzy set into unimodal fuzzy sets has been proposed by Gitman and Levine [42]. The generated partition is optimal in the sense that the clustering procedure detects all of the existing unimodal fuzzy sets and realizes the maximum separation among them. The order of the objects according to their membership degrees as well as the order according to a distance are used in the algorithm. The algorithm is a systematic procedure which always terminates.

Second, Couturier and Fioleau [26] have proposed a method which generates a fuzzy coverage of the set of objects $X = \{x_1, \ldots, x_n\}$. The method is based on the concept of a stable set internally maximum. The cluster is the stable set internally maximum when a representativeness constraint and a separability constraint are satisfied. Coverage of the set of objects by crisp clusters is constructedin the first stage, while values of the membership function are assigned to each element $x_i \in X$ in the second stage of the algorithm. So, the fuzzy coverage $C(X) = \{A^l \mid l = \overline{1,c}, c \leq n\}$ of the initial set of objects $X = \{x_1, \ldots, x_n\}$ is the classification result that is obtained from the algorithm.

Third, an algorithm of Chiang, Yue, and Yin[19] is very good illustration of a heuristic method of fuzzy clustering. Their FCC-algorithm is based on the concept of a fuzzy cover and an objective function meant to group the objects into appropriate fuzzy clusters. The fuzzy cover is defined in [19] as a fuzzy set in which the fuzzy relation of any enclosed object to the centroid object is larger than a certain level. The objective function is incorporated into the FCC-algorithm and this objective function reflects the natural grouping of fuzzy covers. The derived centroids of the fuzzy covers form the backbones of the final clusters after splitting the fuzzy covers into groups and the number of the final clusters does not have to be given a priori because it is determined by an appointed fuzzy relation.

Fourth, a mountain clustering method (MCM) is often considered in the framework of fuzzy clustering and the well-known MCM-algorithm was proposed by Yager and Filev [163]. The algorithm is a simple and efficient clustering procedure for the estimation of cluster prototypes. Let $X = \{x_1, \ldots, x_n\}$ be a collection of n objects in the m_1-dimensional Euclidean space $I^{m_1}(X)$ and the m_1-dimensional hypercube be $I^1 \times I^2 \times \ldots \times I^{m_1}$ where each interval I^{t_1}, $t_1 = 1, \ldots, m_1$, is defined by the range of the coordinate $x_i^{t_1}$, $i = 1, \ldots, n$. So, the hypercube contains all objects $x_i \in X$. Each interval I^{t_1} is subdivided into r^{t_1} equidistant points and the discretization forms a m_1-dimensional grid in the hypercube. This results in $\hat{c} = r^1 \times r^2 \times \ldots \times r^{m_1}$ grid nodes $\breve{\tau}^{\hat{l}}$, $\hat{l} = 1, \ldots, \hat{c}$. The grid nodes are possible centers of clusters A^l, $l \in \{1, \ldots, c\}$. If $d(x_i, \tau^{\hat{l}})$ denotes the distance between $x_i \in X$ and the grid node τ^l, then a mountain function is defined for each node τ^l as

$$M(\tau^{\hat{l}}) = \sum_{i=1}^{n} \exp(-\eta \cdot d(x_i, \tau^{\hat{l}})), \tag{1.109}$$

where $\eta > 0$ is a constant. A node with many neighborhood objects will therefore have a large mountain function value and such a node is more suited to be a potential cluster center. The grid node with the maximum mountain function value $M^*_{(1)}$ in the first step of the process is selected as the first cluster center, $\tau^{(1)}$. Otherwise, if there exists more than one node with the same maximum value of the mountain function, then one node can be selected randomly. The detected cluster center must be removed before obtaining the next cluster center. So, the mountain function, after eliminating the b-th cluster center, is defined by

$$M_{(b+1)}(\tau^{\hat{l}}) = \max\left(0, M_{(b)}(\tau^{\hat{l}}) - M^*_{(b)} \cdot \exp\left(-\zeta \cdot d(\tau^{(b)}, \tau^{\hat{l}})\right)\right), \tag{1.110}$$

where $\zeta > 0$ is a constant, $M_{(b+1)}(\tau^{\hat{l}})$ is the revised mountain function, $M_{(b)}(\tau^{\hat{l}})$ is the previous value of the mountain function, $M^*_{(b)}$ is the current maximum and $\tau^{(b)}$ is the last discovered center. The process is stopped when the condition $\left(M^*_{(b+1)} / M^*_{(1)}\right) < \varepsilon$ is met, where $\varepsilon < 1$ is a positive constant. Note that the derivation of proper values of the parameters η, ζ, and ε is a serious problem. The accuracy of the MCM-algorithm and its computational cost depend on the fitness of the grid. Moreover, with the increase of the dimensionality of the given data set X the computational cost of the MCM-algorithm also increases.

The mountain clustering method was extended by other researchers. For example, a subtractive clustering method (SCM) is a modification of the mountain method and the SCM-algorithm was proposed by Chiu [22]. In this approach, each object $x_i \in X$ is considered as a potential cluster center, τ. So, $\hat{c} = n$ and the mountain function is calculated for each object. Hence, the mountain function is presented by the equation

$$M(x_i) = \sum_{i=1}^{n} \exp(-\eta \cdot d(x_i, x_j)), \tag{1.111}$$

and the discounting of the mountain function is rewritten as

$$M_{(b+1)}(x_i) = \max\left(0, M_{(b)}(x_i) - M^*_{(b)} \cdot \exp\left(-\zeta \cdot d(x_i, \tau^{(b)})\right)\right) \tag{1.112}$$

where $\tau^{(b)}$ is the previously detected center. The number of potential cluster centers depends on the number of objects in the data set X, not on the dimensionality of the data set. On the other hand, the SCM-algorithm will give a good clustering result only if the desired cluster centers is close to one of the objects.

Some other modifications of the mountain clustering method and the subtractive clustering method are described in the literature. Firstly, a generalization of the mountain method for circular shells detection was proposed by Pal and Chakraborty [87], where the corresponding MCS-algorithm was also described. Secondly, the MMCA-algorithm was proposed by Yang and Wu [167]. Thirdly, the WMSC-algorithm was proposed by Chen, Qin and Jia [18].

Finally, it is should be noted, that the heuristic algorithms of fuzzy clustering are useful tools for the exploratory data analysis.

1.2.3 Hierarchical Methods of Fuzzy Clustering

In general, a hierarchy of clusters is a result of the application of some hierarchical fuzzy clustering method to a set of objects $X = \{x_1, \ldots, x_n\}$. The hierarchy of clusters can be constructed using different techniques. Let us consider in brief the hierarchical methods of fuzzy clustering.

First, a method of hierarchical classification based on the (max-min) transitive closure $\tilde{T}$ of a fuzzy tolerance T has been proposed by Tamura, Higuchi, and Tanaka [106]. So, the (max-min) transitive closure $\tilde{T}$ generates a hierarchical classification of the set of objects $X = \{x_1, \ldots, x_n\}$ and a crisp partition can be induced by the α-level set $\tilde{T}_{\alpha_\ell}$ of the fuzzy equivalence relation $\tilde{T}$ for some value $\alpha_\ell \in (0,1]$.

Second, a heuristic method for constructing the estimated fuzzy transitive relation has been proposed by Watada, Tanaka, and Asai [154] and the method is a basis for a hierarchical clustering algorithm. The deviation between the estimated transitive and the given intransitive fuzzy relations should be minimized. In other words, let $X = \{x_1, \ldots, x_n\}$ be a set of objects and T be a given fuzzy tolerance relation on X with $\mu_T(x_i, x_j)$ being its membership function. The purpose of the estimation task is to obtain the fuzzy transitive relation $\tilde{T}$ which minimizes

$$Q(\tilde{T}) = \sum_{x_i, x_j \in X} \left(\mu_T(x_i, x_j) - \mu_T(x_i, x_j)\right)^2, \tag{1.113}$$

under the (max-min) transitivity constraints (1.44). So, the problem of classification boils down to the construction of a hierarchy of crisp clusters based on the fuzzy relation $\tilde{T}$.

Third, a fuzzy divisive hierarchical (FDH) clustering method has been proposed by Dumitrescu [34] and a multilevel fuzzy classification is obtained from the corresponding FDH-algorithm. Therefore, each fuzzy partition is a refinement of the fuzzy partition which corresponds to the previous level of classification. Since the FDH-algorithm produces a hierarchy of fuzzy clusters, the clustering procedure is essentially different from both the previously considered hierarchical clustering methods based on fuzzy relations which yield hierarchies of crisp clusters. The optimal number of fuzzy clusters in the data set is an outcome of the FDH-algorithm. That is why no a priori knowledge about the cluster structure of the data set is necessary. This is a sufficient reason to consider that Dumitrescu's clustering procedure also provides a solution to the cluster validity problem. Moreover, the proposed classification method can be applied to the solution of the dimensionality reduction problem and the corresponding FHDR-algorithm is also described in [34].

Fourth, Geva [40] has propose a hierarchical unsupervised fuzzy clustering (HUFC) algorithm and which is based on a weighted version of the GG-algorithm. The HUFC-algorithm includes a recursive call to a two-step procedure in which the relevant features of the data in the current level are extracted and reduced in the first step of the recursive process, and the a modified GG-algorithm is applied to the relevant part of the data in the second step of each recursive call. The number of fuzzy clusters in each step is determined using an adapted validity measure which is based on the fuzzy hypervolume (1.92). The recursive process is terminated when the optimum number of fuzzy clusters is equal one, or when the number of objects in a cluster is smaller than some a priori determined constant multiplied by the number of features. The sum of membership values for each object in all fuzzy clusters is equal to one when the HUFC-algorithm is stopped.

Fifth, an unsupervised fuzzy graph clustering (UFGC) method has been developed by Devillez, Billaudel, and Villermain Lecolier and the UFGC-algorithm is described in [31]. The first stage of the UFGC-algorithm consists in using the FCM-algorithm to divide the set of objects into c' fuzzy sub-clusters, $A^{l'}$, $l'=1,\ldots,c'$. Notice, that the number of fuzzy sub-clusters c' must be greater than the number of real classes c, $c'>c$, and each fuzzy sub-cluster must belong to one real class only. The construction of a neighborhood matrix, which is the matrix of proximity values between fuzzy sub-clusters, should be made in the second stage of the UFGC-algorithm. Proximity values between fuzzy sub-clusters can be calculated using a similarity degree defined by Frigui and Krishnapuram [38]:

$$\mu_T(A^{l'}, A^{a'}) = 1 - \frac{\sum_{x_i \in A^{l'}} |u'_{l'i} - u'_{a'i}|}{\sum_{x_i \in A^{a'}} u'_{a'i} + \sum_{x_i \in A^{l'}} u'_{l'i}}, \ l', a' = 1, \ldots, c', \tag{1.114}$$

where the element $u'_{l'i}$ represents the membership grade of object $x_i \in X$ to the fuzzy sub-cluster $A^{l'}$, $l' \in \{1, \ldots, c'\}$ and the element $\mu_T(A^{l'}, A^{a'})$ corresponds to the similarity value between the fuzzy sub-clusters $A^{l'}$ and $A^{a'}$, $l', a' \in \{1, \ldots, c'\}$. Thus, the matrix $T_{c' \times c'} = [\mu_T(A^{l'}, A^{a'})]$ of a fuzzy tolerance is obtained and it defines a fuzzy proximity graph in which the vertices represent the fuzzy sub-clusters and the arcs represent the links. Moreover, it is possible to define a remoteness matrix. Therefore, a tree of graduated hierarchy can be associated to the fuzzy proximity graph and it is graduated according to the remoteness values. To recover the real classes it is necessary to cut the tree of graduated hierarchy at an appropriate level. This operation corresponds to the last stage of the UFGC-algorithm which consists in searching for the connected components of the proximity graph. For this purpose, a validity measure is also proposed in [31], and the minimum value of the proposed validity measure, as a function of the number of classes, is sought which is attained for the number c of real classes and corresponds to the appropriate level value of the graduated hierarchy.

It is should be noted that a similar idea has been used by Dong, Zhuang, Chen, and Tai [32] for constructing their hierarchical FHC-algorithm based on the connectedness of a fuzzy graph. However, the threshold value of the level for the fuzzy graph should be given a priori.

Sixth, the hierarchical FCM-algorithm has been proposed by Pedrycz and Reformat [92] which is based on using the method of conditional fuzzy c -means proposed by Pedrycz [90]. The CFCM-algorithm is based on an iterative optimization of the fuzzy objective function (1.72) under the constraints

$$u_{li} \geq 0, \ \sum_{l=1}^{c} u_{li} = f_i, \ i = 1, \ldots, n, \ l = 1, \ldots, c, \tag{1.116}$$

where $f_i \in [0,1]$, $i = 1, \ldots, n$ describes the level of involvement of object $x_i \in X$ in the constructed fuzzy clusters A^l, $l = 1, \ldots, c$. So, the FCM-algorithm is applied to the set of objects $X = \{x_1, \ldots, x_n\}$ in the first step of the HFCM-algorithm and the CFCM-algorithm is applied to the obtained fuzzy clusters in the following steps. In other words, the FCM-algorithm forms the first layer and the CFCM-algorithm forms all next layers of the hierarchy. That is why the HFCM-algorithm can be considered as a fuzzy divisive hierarchical clustering procedure.

Seventh, a fuzzy agglomerative hierarchical clustering (FAHC) method has been proposed by Horng, Chen, Chang, and Lee [50] which uses the similarity threshold value and the difference threshold value.

It is should be noted that the hierarchical algorithms of fuzzy clustering are very useful for a detaile analysis of the set of objects. However, the fuzzy hierarchical clustering methods require a great memory space when the number of objects increases which can make their applications difficult in nontrivial real problems.

1.3 Methods of Possibilistic Clustering

The conditions (1.71) for a fuzzy c -partition are very difficult to fulfill for essential reasons. Moreover, the constraints possibly lead to undesirable membership degrees of some data. Let us assume that an object $x_i \in X$ exists which is in a great distance to all compact clusters in the data set. This object can be interpreted as an outlier and as such it would be assigned approximately the same membership degree $u_{li} = 1/c$ to all fuzzy clusters A^l, $l = 1,\ldots,c$. The effect can be illustrated by the following simple example. Let $X = \{x_1,\ldots,x_{16}\}$ be the set of objects which were considered by Yang and Wu [168]. The corresponding two-dimensional data set is presented in Figure 1.6.

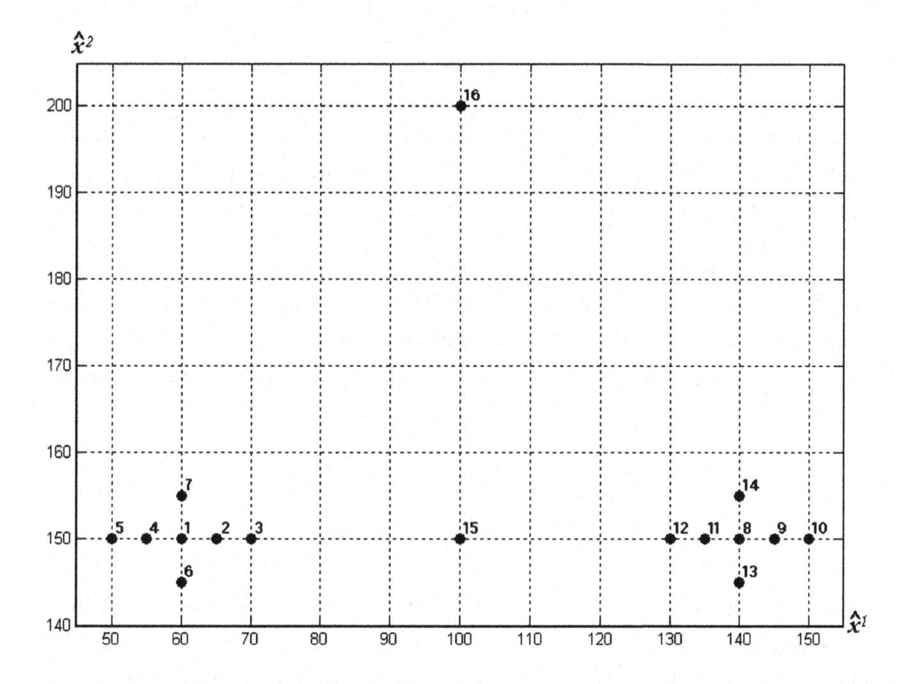

Fig. 1.6 Two clusters' data set with a bridge object and an outlier

In Figure 1.6 there are two clusters with a bridging object x_{15} and an outlying object x_{16}. The data set was processed by the FCM-algorithm with the number of fuzzy clusters $c = 2$ and the weighting exponent value $\gamma = 2$. The membership functions of two fuzzy clusters are shown in Figure 1.7. The membership values of the first class are represented by ○ andhe membership values of the second class are represented by ■.

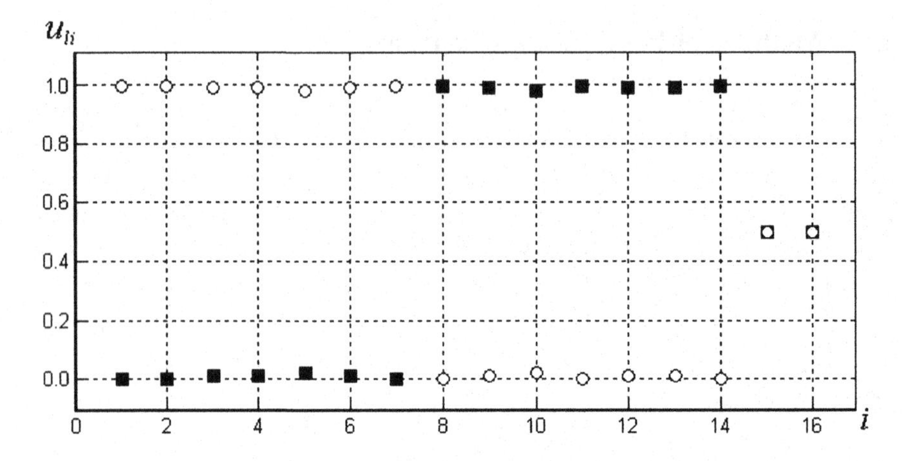

Fig. 1.7 Membership functions of two fuzzy clusters obtained from the FCM-algorithm

The FCM-algorithm assigns the membership 0.5 in the two fuzzy clusters to both x_{15} and x_{16}. However, the object x_{16} should not have a high value of the membership degree in either of the fuzzy cluster. Therefore, the membership value 0.5 is employed.

To avoid such a drawback anapproach topossibilistic clustering has been proposed by Krishnapuram and Keller [66]. The concept of a possibilistic partition is a basis of hat possibilistic clustering methods. For each object x_i, $i = 1,\ldots,n$ the grades of membership should satisfy the conditions of a possibilistic partition:

$$\sum_{l=1}^{c} \mu_{li} > 0, \; 0 \leq \mu_{li} \leq 1, \; 1 \leq l \leq c. \tag{1.117}$$

The family of fuzzy sets $Y(X) = \{A^l \mid l = \overline{1,c}, c \leq n\}$ is a possibilistic partition of the initial set of objects $X = \{x_1,\ldots, x_n\}$ if condition (1.117) is met. With these constraint the membership values μ_{li}, $i = 1,\ldots,n$, $l = 1,\ldots,c$, could be interpreted as the values of a typicality degree. In other words, the membership degree μ_{li} is interpreted as the degree of representativeness of an object $x_i \in X$

for a fuzzy cluster $A^l \in Y(X)$. On the other hand, the membership degree μ_{li} is equal to the value of a possibility distribution function for cluster A^l over the universe consisting of all objects x_i. This possibility distribution will be denoted by $\pi_l(x_i)$.

The possibilistic approach to clustering was developed by other researchers too. This approach can be considered as an example of an optimization approach in fuzzy clustering because all methods of possibilistic clustering are objective function-based. Similarly to the fuzzy clustering procedures, the possibilistic clustering algorithms can be divided into two types: object type and relational type. Moreover, hybrid clustering techniques have also been proposed which combine the fuzzy and possibilistic approaches. That is why the hybrid clustering techniques produce probabilistic memberships in the sense of (1.71) and possibilistic memberships in the sense of (1.117) simultaneously.

1.3.1 A Brief Review of Prototype-Based Methods

Let us consider some possibilistic clustering algorithms, and we will consider firstthe object data clustering methods. A possibilistic generalization of the FCM-algorithm, or PCM for short, has been proposed by Krishnapuram and Keller [66] and this is the first possibilistic clustering procedure. The objective function of this possibilistic clustering model can be written as follows:

$$Q_{PCM_1}(Y,\mathrm{T}) = \sum_{l=1}^{c}\sum_{i=1}^{n} \mu_{li}^{\psi} d^2(x_i, \tau^l) + \sum_{l=1}^{c} \eta_l \sum_{i=1}^{n} (1-\mu_{li})^{\psi}, \tag{1.118}$$

where μ_{li}, $l = 1,\ldots,c$, $i = 1,\ldots,n$, are the values of a typicality degree, x_i, $i \in \{1,\ldots,n\}$, is an element of the data set, $\mathrm{T} = \{\tau^1,\ldots,\tau^c\}$ is the set of fuzzy clusters prototypes, $d^2(x_i, \tau^l)$ is the squared Euclidean distance between x_i and τ^l (1.73), the parameter $\eta_l > 0$ determines the permissible extension of fuzzy cluster A^l, $l = 1,\ldots,c$, and $\psi > 1$ is a weighting exponent for the possibilistic memberships. The value of ψ determines the shape of the possibility distribution $\pi_l(x_i)$. In most cases the value for the fuzzifier ψ between 1 and 2 will be appropriate.

The first term of the objective function (1.118) is equal to the objective function (1.72) whereas the second term forces μ_{li} to be as large as possible to avoid a trivial solutions. The purpose of the classification task is to obtain solutions $Y(X)$ and $\tau^1,\ldots,\tau^c$ which minimize (1.118). The scale parameter η_l of a cluster corresponds to the size of a cluster and this parameter has a major influence on the clustering results. If the cluster shapes are known in advance, η_l

could easily be estimated for all $l = 1,\ldots,c$. For example, when the initial data set is relatively noise-free, the FCM-algorithm can be used to obtain (initialize) the initial possibilistic partition while running the PCM-algorithm, and to obtain an appropriate estimate of η_l. Krishnapuram and Keller [66] recommend to select η_l as

$$\eta_l = K \frac{\sum_{i=1}^{n} \mu_{li}^{\psi} d^2(x_i, \tau^l)}{\sum_{i=1}^{n} \mu_{li}^{\psi}}, \tag{1.119}$$

where $K \in (0,\infty)$ is typically chosen to be one. Another method to estimate the parameter η_l for all $l = 1,\ldots,c$ has also been proposed in [67]. The value of the parameter η_l can be fixed for all iterations of the PCM-algorithm or it may vary in each iteration.

The criterion (1.118) is not unique an objective function for the PCM-algorithm. Krishnapuram and Keller [67] give as an alternative objective function:

$$Q_{PCM_2}(Y,\mathrm{T}) = \sum_{l=1}^{c}\sum_{i=1}^{n} \mu_{li} d^2(x_i, \tau^l) + \sum_{l=1}^{c} \eta_l \sum_{i=1}^{n} (\mu_{li} \log \mu_{li} - \mu_{li}). \tag{1.120}$$

Note that the first term of the objective function (1.120) is equal to the first term of the objective function (1.79). Compared with objective function (1.118), the objective function (1.120) does not involve a weighting exponent for the possibilistic memberships. So, it is not necessary to select a value for this parameter. Note that $(\mu_{li} \log \mu_{li} - \mu_{li})$ in the objective function (1.120) is a monotone decreasing function in $[0,1]$, similar to $(1-\mu_{li})^{\psi}$ in the objective function (1.118).

The application of the PCM-algorithm can be illustrated on a simple example of Yang and Wu's [168] data set. The two-dimensional data set is shown in Figure 1.6. The data set was processed by the PCM-algorithm based on the objective function (1.120) with the number of fuzzy clusters $c = 2$ and the weighting exponent value $\psi = 2$. The PCM-algorithm was performed with the initialization of cluster centers: $\tau^1 = (60, 150)$ and $\tau^2 = (145, 150)$. Thus, the objects x_1 and x_9 were selected as the initialization of cluster centres. The possibilistic memberships μ_{li}, $l = 1,2$, $i = 1,\ldots,16$ of two clusters are shown in Figure 1.8 and the values which equal to zero are not shown in the figure. The membership values of the first class are represented by ○ and the membership values of the second class are represented by ■.

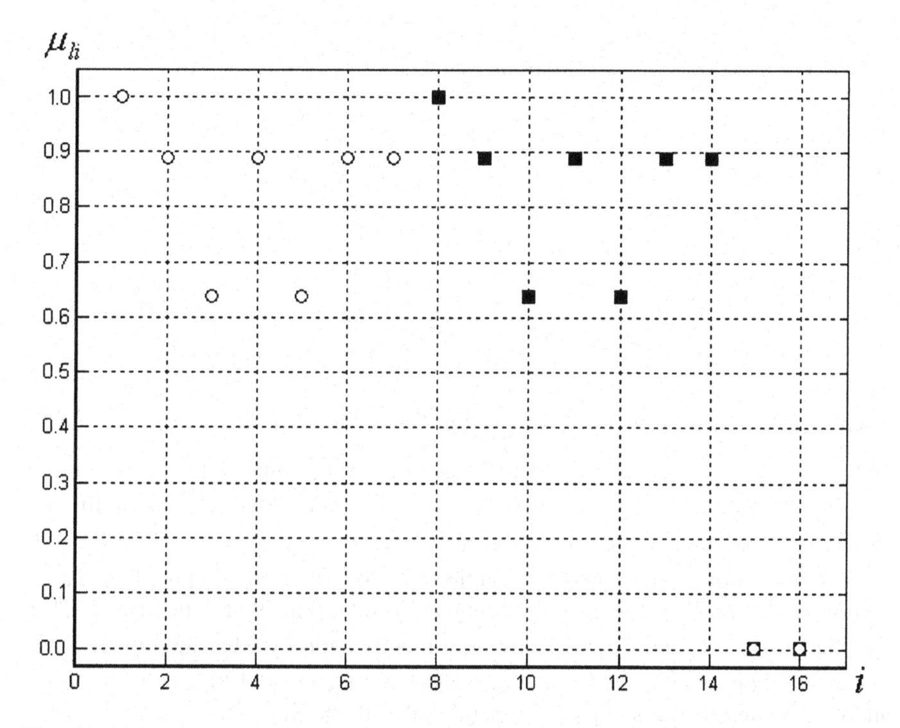

Fig. 1.8 Possibilistic membership values of two clusters obtained by the PCM-algorithm based on the objective function (1.120) for Yang and Wu's data set

The PCM-algorithm is not a unique clustering procedure which has been proposed by Krishnapuram and Keller in the framework of the possibilistic approach to clustering. Possibilistic analogs of other fuzzy objective functions are also considered in [66] where the corresponding PGK-algorithm and PCSS-algorithm are proposed.

A possibilistic clustering method based on a robust approach using Vapnik's [109] ε -intensive estimator, called as the εPCM-algorithm, has been proposed by Łęski [74] . The ε -intensive loss function is:

$$\left|t\right|_{\varepsilon} = \begin{cases} 0, & |t| \le \varepsilon \\ |t| - \varepsilon, & |t| > \varepsilon \end{cases}, \tag{1.121}$$

where t denotes the value of error and ε denotes the value of a intensivity parameter. Many robust loss functions are described in literature. However, due to its simplicity Vapnik's ε -intensive loss function (1.121) is of special interest. For example, the well-known absolute error loss function is a special case of (1.121) for $\varepsilon = 0$.

If the ε -intensive loss function (1.121) is put into the possibilstic c -means criterion (1.118), then the objective function takes the form [74]:

$$Q_{\varepsilon PCM}(Y,\mathrm{T}) = \sum_{l=1}^{c}\sum_{i=1}^{n}\mu_{li}^{\psi}\left|x_i - \tau^l\right|_{\varepsilon} + \sum_{l=1}^{c}\eta_l\sum_{i=1}^{n}(1-\mu_{li})^{\psi}, \tag{1.122}$$

where

$$\left|x_i - \tau^l\right|_{\varepsilon} = \sum_{t=1}^{m}\left|x_i^t - \tau^{(t)l}\right|_{\varepsilon}. \tag{1.123}$$

Therefore, the purpose of the classification task is to obtain the solutions $Y(X)$ and $\tau^1,\ldots,\tau^c$ which minimize (1.122) under the constraints (1.117). A test on a real-world database shows a small sensitivity of the εPCM-algorithm to the election of the parameters ε and η_l [74].

A modification of the objective function (1.120) has been proposed by Yang and Wu [168] where the corresponding PCA-algorithm is also described. The parameter η_l is replased in their proposed modification by other multiplier. That is why the proposed modification of the objective function (1.120) can be properly analyzed.Yang and Wu's objective function takes the form:

$$Q_{PCA}(Y,\mathrm{T}) = \sum_{l=1}^{c}\sum_{i=1}^{n}\mu_{li}^{\psi}d^2(x_i,\tau^l) + \frac{\zeta}{\psi^2\sqrt{c}}\sum_{l=1}^{c}\sum_{i=1}^{n}(\mu_{li}^{\psi}\log\mu_{li}^{\psi} - \mu_{li}^{\psi}), \tag{1.124}$$

where $\zeta = \frac{1}{n}\sum_{i=1}^{n}\|x_i - \overline{x}\|$ with $\overline{x} = \frac{1}{n}\sum_{i=1}^{n}x_i$ and $d^2(x_i,\tau^l)$ is the squared Euclidean distance (1.73). The parameter ζ is the normalization term that measures the grade of separation of the data set and this parameter can always be fixed as the sample co-variance. The parameter $\sqrt{c}$ is used in (1.124) to control the steepness degree of the membership functions. The role of the parameter ψ is equal to the fuzzifier in the objective function (1.118).

Yang and Wu note in [168] that the purpose of the objective function (1.124) is to find prototypes such that the sum of their membership function be maximized. The results that maximize the partition coefficient (1.89) will be obtained from the PCA-algorithm in the case when the weighting exponent value $\psi = 2$. Moreover, the minimization of the objective function (1.124) is equivalent to to maximization of the mountain function in the MCM-algorithm. That is why the cluster prototypes obtained from the PCA-algorithm correspond to to the peaks of the mountain function.

An example of the PCA-algorithm performance can be illustrated by using Yang and Wu's data set presented in Figure 1.6 in which the objects x_1 and x_9 are selected as the initialization of the cluster centres. The data set is processed by the PCA-algorithm with the number of fuzzy clusters $c = 2$ and the weighting exponent value $\psi = 2$. The possibilistic membership values of two clusters are shown in Figure 1.9 where the membership values of the first class are represented by ○ and the membership values of the second class are represented by ■. The values of the typicality degree which equal zero are not shown in this figure.

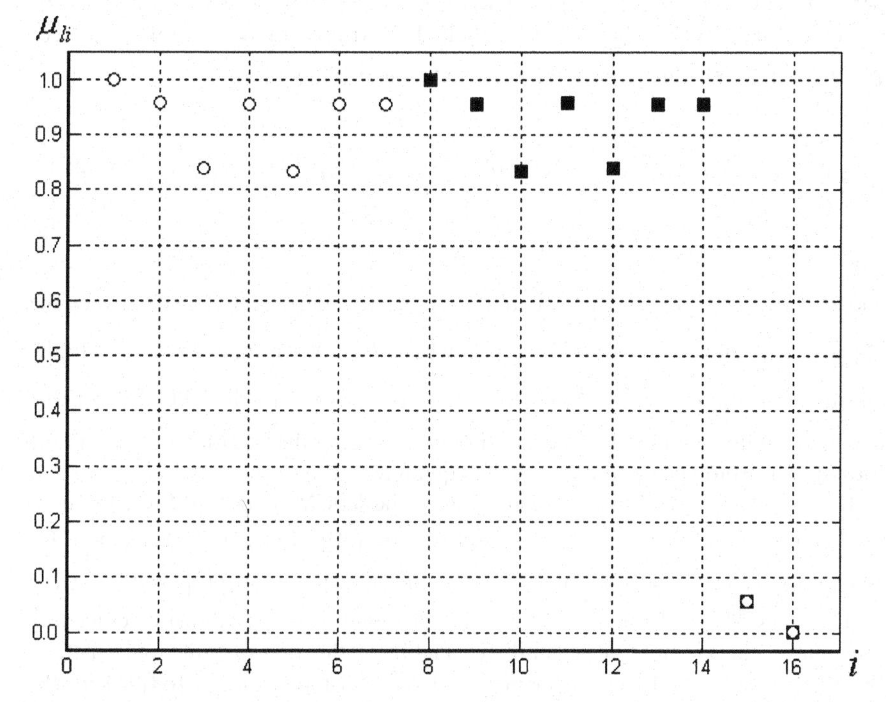

Fig. 1.9 Possibilistic membership values of two clusters obtained from the PCA-algorithm for Yang and Wu's data set

Therefore, equal possibilistic membership values are obtained from the PCM-algorithm based on the objective function (1.120) for the bridge object x_{15} and the outlying object x_{16} and different possibilistic membership values are obtained from the PCA-algorithm for these objects. Note that the maximal values of the typicality degree are obtained in both numerical experiments for the objects x_1 and x_8.

Some other possibilistic clustering procedures have also been developed. In particular, the EPCM-algorithm of possibilistic clustering has been proposed by Xie, Wang, and Chung [159] and the PGCM-algorithm has been introduced by

Ménard, Courboulay, Dardignac [77]. Moreover, many interesting results have been obtained by other authors and presented in numerous papers.

1.3.2 A Relational Algorithm of Possibilistic Clustering

A relational algorithm that is in a sense dual to the PCM-algorithm has been given by Hathaway, Bezdek, and Davenport [46] where relational versions of the c-means algorithms have been reviewed. A relational approach to possibilistic clustering can be explained by the relational possibilistic clustering method (RPCM) which has been proposed by De Cáceres, Oliva and Font [30]. The objective function (1.118) is used for their RPCM-algorithm where the distances $d^2(x_i, \tau^l)$ to the cluster prototypes are calculated using a formula

$$d^2(x_i, \tau^l) = \left(\sum_{i=1}^{n} \mu_{li}^{\psi} d^2(x_i, x_j) \right) \Big/ \left(\sum_{i=1}^{n} \mu_{li}^{\psi} \right) - (1/2) \cdot \left(\sum_{i,j=1}^{n} \mu_{li}^{\psi} \mu_{lj}^{\psi} d^2(x_i, x_j) \right) \Big/ \left(\sum_{i=1}^{n} \mu_{li}^{\psi} \right)^2, \tag{1.125}$$

and $d^2(x_i, x_j)$ represents the squared dissimilarity coefficient values between the objects x_i and x_j, $i, j = 1, \ldots, n$, where $d(x_i, x_j)$ is an element of the dissimilarity relation matrix $\rho_{n \times n} = [d(x_i, x_j)]$. That is why the RPCM-algorithm can be obtained from the PCM-algorithm by replacing the equation (1.73) and the formula for prototype updating with the equation (1.125).

It is should be noted that the extension of the RPCM-algorithm for some non-Euclidean relation $\rho_{n \times n}$ is also discussed in [30]. Moreover, a method for estimating the parameters η_l on based on cluster variability is proposed in [30].

The RPCM-algorithm is not a unique relational possibilistic clustering procedure. For example, a possibilistic analog of the fuzzy c-medoids method (FCMdd) is discussed by Krishnapuram, Joshi, Nasraoui and Yi [68] where the FCMdd-algorithm and some of its possible modifications are proposed.

1.3.3 Hybrid Clustering Techniques

Hybrid clustering techniques generate the membership valuess u_{li} and typicality values μ_{li} for all $l = 1, \ldots, c$, $i = 1, \ldots, n$. Such an approach has been proposed by Pal, Pal, and Bezdek [85] where a fuzzy-possibilistic c-means (FPCM) algorithmhas been proposed. The FPCM-algorithm constrains the typicality values so that the sum of them over all objects $x_i \in X$ to a cluster is equal one. In other words, the conditions of the fuzzy c-partition (1.71) are met for the possibilistic membership values μ_{li}, $l = 1, \ldots, c$, $i = 1, \ldots, n$. As for typicality values obtained from the FPCM-algorithm, one may need to scale them up. To overcome this problem, Pal, Pal, Keller and Bezdek [86] proposed the PFCM-algorithm so that the constraints (1.117) are met for the typicality values. The PFCM-algorithm is based on the objective function:

$$Q_{PFCM}(Y,\mathrm{T}) = \sum_{l=1}^{c}\sum_{i=1}^{n}(au_{li}^{\gamma} + b\mu_{li}^{\psi})d^2(x_i,\tau^l) + \sum_{l=1}^{c}\eta_l\sum_{i=1}^{n}(1-\mu_{li})^{\psi}, \quad (1.126)$$

where u_{li} is the probabilistic membership of x_i in the cluster A^l and it fulfills the constraints (1.71), μ_{li} is the possibilistic membership of x_i in the cluster A^l and it fulfills the constraints (1.117), $\gamma > 1$ is the weighting exponent for the probabilistic membership value, $\psi > 1$ is the weighting exponent for the possibilistic membership value, $d^2(x_i,\tau^l)$ is the squared Euclidean distance (1.73), the parameters $\eta_l > 0$ are user defined constants and the constants $a > 0$ and $b > 0$ define the relative importance of the probabilistic and possibilistic membership values in the objective function (1.126).

Note that in (1.126) the setting of the constraint $a + b = 1$ is not expected to result in good prototypes. If $b = 0$ and $\eta_l = 0$ for all A^l, $l = 1,\ldots,c$, then (1.126) reduces to the objective function (1.72); while $a = 0$ and $b = 1$ convert it to the objective function (1.118).

An improvement of the PCM-algorithm (IPCM) has been proposed by Zhang and Leung [174] and is based on the generalizations of the criteria (1.118) and (1.120).

First, the generalization of the objective function (1.118) for the IPCM-algorithm can be written as follows:

$$Q_{IPCM_1}(P,Y,\mathrm{T}) = \sum_{l=1}^{c}\sum_{i=1}^{n}u_{li}^{\gamma}\mu_{li}^{\psi}d^2(x_i,\tau^l) + \sum_{l=1}^{c}\eta_l\sum_{i=1}^{n}u_{li}^{\gamma}(1-\mu_{li})^{\psi}, \quad (1.127)$$

where $d^2(x_i,\tau^l)$ is the squared Euclidean distance (1.73) between x_i and τ^l, and the parameter η_l in (1.127) is generalized in the following way:

$$\eta_l = \frac{\sum_{i=1}^{n}\mu_{li}^{\psi}u_{li}^{\gamma}d^2(x_i,\tau^l)}{\sum_{i=1}^{n}\mu_{li}^{\psi}u_{li}^{\gamma}}. \quad (1.128)$$

In other words, the possibilistic and probabilistic membership values are incorporated in the scale parameter η_l, $l = 1,\ldots,c$. Note that the parameter η_l involves all distances including those of the outliers.

Second, the generalization of the objective function (1.120) for the IPCM-algorithm takes the form:

$$Q_{IPCM_2}(P,Y,\mathrm{T}) = \sum_{l=1}^{c}\sum_{i=1}^{n} u_{li}^{\gamma} \mu_{li} d^2(x_i,\tau^l) + \sum_{l=1}^{c} \eta_l \sum_{i=1}^{n} u_{li}^{\gamma} (\mu_{li} \log \mu_{li} - \mu_{li} + 1), \tag{1.129}$$

It should be noted that the weighting exponent $\psi > 1$ for the possibilistic membership value μ_{li} is omitted in (1.129) but the weighting exponent $\gamma > 1$ for the probabilistic membership value u_{li} is present in the objective function.

An unsupervised possibilistic fuzzy clustering (UPFC) method has been elaborated by Wu, Wu, Sun and Fu [157], and its ideas is to minimize the following objective function:

$$Q_{UPFC}(P,Y,\mathrm{T}) = \sum_{l=1}^{c}\sum_{i=1}^{n} (a u_{li}^{\gamma} + b \mu_{li}^{\psi}) d^2(x_i,\tau^l) + \frac{\zeta}{\psi^2 \sqrt{c}} \sum_{l=1}^{c}\sum_{i=1}^{n} (\mu_{li}^{\psi} \log \mu_{li}^{\psi} - \mu_{li}^{\psi}), \tag{1.130}$$

subject to the constraints (1.71) and (1.117), where $d^2(x_i,\tau^l)$ is the squared Euclidean distance (1.73). The first term of the objective function (1.130) is equal to the first term of the objective function (1.126) and the second term of the objective function (1.130) is equal to thesecond term of the objective function (1.124). That is why the constants $a > 0$ and $b > 0$ in the objective function (1.130) have the same meaning as those in the objective function (1.126). Similarly, the parameter ζ in the UPFC-algorithm has the same interpretation as in the PCA-algorithm.

A numerical experiment can be performed on Yang and Wu's data set. The computational setup is as follows: the number of fuzzy clusters is $c = 2$, the value of the weighting exponent for the probabilistic membership values is $\gamma = 2$, the value of the weighting exponent for the possibilistic membership values is $\psi = 2$, the values of the constants a and b are equal 1, and the objects x_1 and x_9 are selected as the initialization of the cluster centres. The values of the probabilistic membership values obtained by the UPFC-algorithm are shown in Figure 1.10 where the membership values of the first class are represented by ○ and the membership values of the second class are represented by ■.

So, the probabilistic membership values obtained from the UPFC-algorithm are similar to the values obtained from the FCM-algorithm. The values of the typicality degree obtained from the UPFC-algorithm are shown in Figure 1.11 where the values of the first class are represented by ○ and values of the second class are represented by ■; the values which equal zero are not shown in this figure.

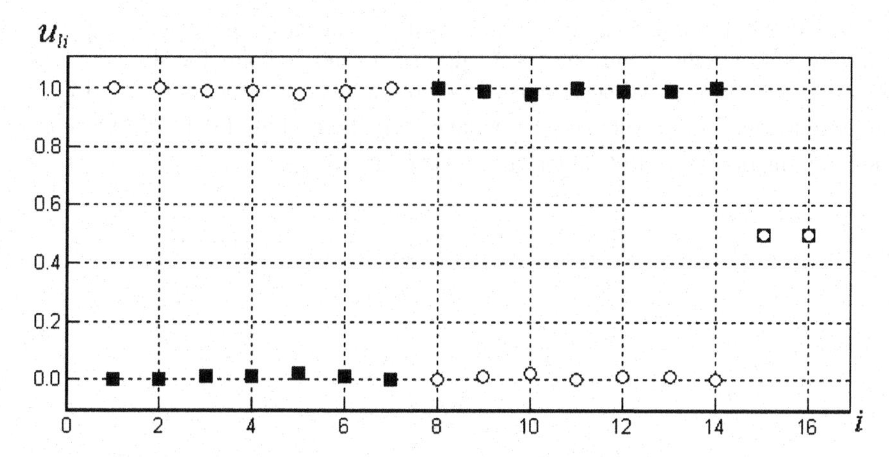

Fig. 1.10 Probabilistic membership values of two clusters obtained from the UPFC-algorithm for Yang and Wu's data set

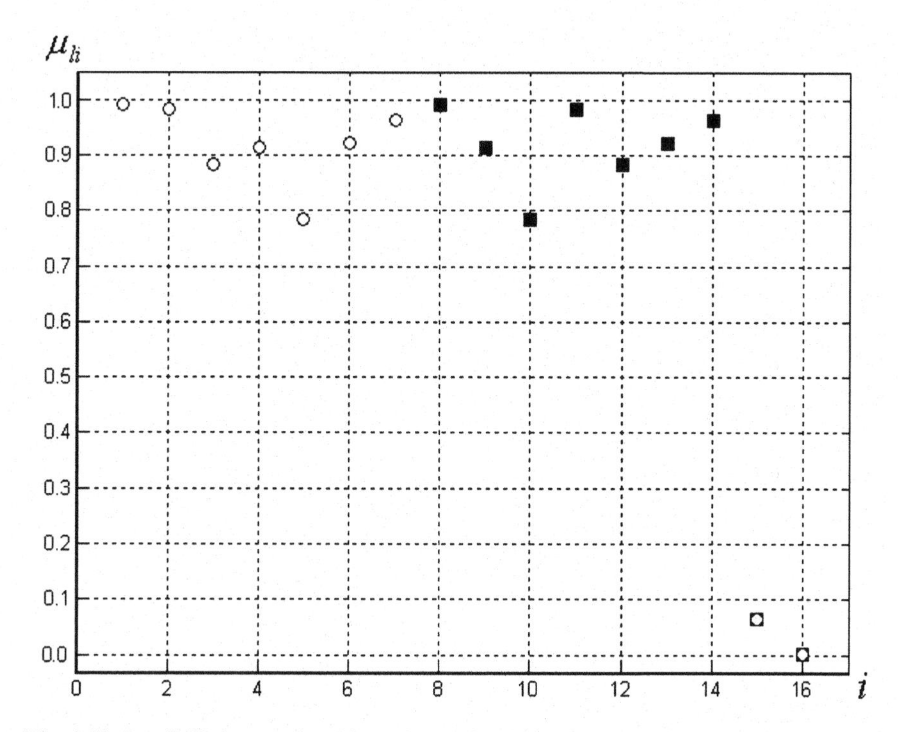

Fig. 1.11 Possibilistic membership values of two clusters obtained from the UPFC-algorithm for Yang and Wu's data set

The PCM-algorithm and the PCA-algorithm can easily generate coincident clusters. On the other hand, the UPFC-algorithm can overcome the shortcomings of coincident clusters and noise sensitivity. Furthermore, as opposed to the PFCM-algorithm which must calculate the parameter η_l by running the FCM-algorithm until termination, the UPFC-algorithm need not do that.

Chapter 2
Heuristic Algorithms of Possibilistic Clustering

As Ruspini [99] has noted, fuzzy clustering is a technique of representation of the initial set of objects by fuzzy clusters. The structure of the set of objects can be described by some fuzzy tolerance relation. A fuzzy cluster can be understood as some fuzzy subset implied by a fuzzy tolerance relation stipulating that a similarity degree of the elements of the fuzzy subset is not less than some threshold value. The condition stated above is a conceptual basis for the proposed approach to clustering which will be described in detail in this chapter.

2.1 Outline of the Approach

This section will include parts which will step by step present our proposed approach. The first subsection includes basic definitions. Illustrative examples of these definitions are considered in the second subsection. The third subsection of the section provides some details on an important problem of data preprocessing.

2.1.1 Basic Definitions

The very essence of the problem formulation of cluster analysis under fuzziness, as meant in this book, can be defined in general as the problem of finding a unique representation of the initial set of objects by fuzzy clusters in the sense of a given definition of what is meants by the concept of a fuzzy cluster. The concept of a representation hs been used in [111] where a man-machine approach to fuzzy classification is described. However, the concept of a representation has a specific meaning in pattern recognition. That is why the term of *allotment* among fuzzy clusters has been used in [114] and in the following publications.

Let us remind the basic concepts of the clustering method based on the concept of allotment among fuzzy clusters. The concept of a fuzzy tolerance relation is the basis for the concept of a fuzzy α-cluster. Let $X = \{x_1, \ldots, x_n\}$ be the initial set of elements and $T : X \times X \to [0,1]$ be some binary fuzzy tolerance relation on $X = \{x_1, \ldots, x_n\}$ with $\mu_T(x_i, x_j) \in [0,1]$, $\forall x_i, x_j \in X$, being its membership function. The essence of the method considered here does not depend on a particular kind of the fuzzy tolerance relation. That is why the method is described

D.A. Viattchenin: *A Heuristic Approach to Possibilistic Clustering*, Studfuzz 297, pp. 59 – 118.
DOI: 10.1007/978-3-642-35536-3_2

here for any fuzzy tolerance relation T. A fuzzy tolerance relation T on $X = \{x_1,...,x_n\}$ can clearly be represented by a matrix $T_{n \times n} = [\mu_T(x_i, x_j)]$, $i, j = 1,\ldots,n$.

The number c of fuzzy clusters can be equal to the number of objects, n. This is taken into account in our further considerations. Let $X = \{x_1,...,x_n\}$ be the initial set of objects. Let T be a fuzzy tolerance relation on X and α be α-level value of T, $\alpha \in (0,1]$. The columns or lines of the fuzzy tolerance matrix are fuzzy sets $\{A^1,..., A^n\}$. Let $\{A^1,..., A^n\}$ be fuzzy sets on X which are generated by a fuzzy tolerance relation T.

Definition 2.1. *The α-level fuzzy set $A^l_{(\alpha)} = \{(x_i, \mu_{A^l}(x_i)) \mid \mu_{A^l}(x_i) \geq \alpha, l \in [1,n]\}$ is a fuzzy α-cluster or, simply, a fuzzy cluster. So, $A^l_{(\alpha)} \subseteq A^l$, $\alpha \in (0,1]$, $A^l \in \{A^1,\ldots, A^n\}$, and μ_{li} is the membership degree of the element $x_i \in X$ for some fuzzy cluster $A^l_{(\alpha)}$, $\alpha \in (0,1]$, $l \in [1,n]$. The value of α is the tolerance threshold of elements of the fuzzy clusters.*

The membership degree of the element $x_i \in X$ for some fuzzy cluster $A^l_{(\alpha)}$, $\alpha \in (0,1]$, $l \in [1,n]$ can be defined as a

$$\mu_{li} = \begin{cases} \mu_{A^l}(x_i), & x_i \in A^l_\alpha \\ 0, & otherwise \end{cases}, \tag{2.1}$$

where an α-level set $A^l_\alpha = \{x_i \in X \mid \mu_{A^l}(x_i) \geq \alpha\}$, $\alpha \in (0,1]$, of a fuzzy set A^l is the support of the fuzzy cluster $A^l_{(\alpha)}$. So, the condition $A^l_\alpha = Supp(A^l_{(\alpha)})$ is met for each fuzzy cluster $A^l_{(\alpha)}$, $\alpha \in (0,1]$, $l \in \{1,\ldots,n\}$. The value of the membership function of each element of the fuzzy cluster in the sense of Definition 2.1 is the degree of similarity of the object to some typical object of fuzzy cluster. So, the membership degree can be interpreted as a degree of typicality of an element to a fuzzy cluster and the fuzzy clusters in the sense of Definition 2.1 are fundamentally different from the fuzzy clusters in the sense (1.71).

In other words, if the columns or lines of the fuzzy tolerance matrix T are fuzzy sets $\{A^1,..., A^n\}$ on X, then the fuzzy clusters $\{A^1_{(\alpha)},..., A^n_{(\alpha)}\}$ are fuzzy subsets of fuzzy sets $\{A^1,..., A^n\}$ for some value α, $\alpha \in (0,1]$. The value zero for the membership function of a fuzzy set is clearly equivalent to the non-belongingness of the particular element to the fuzzy set. That is why the values of the tolerance threshold α are considered as belonging to the interval $(0,1]$.

Definition 2.2. *Let* T *be a fuzzy tolerance relation on* X, *where* X *is the set of elements, and* $\{A^1_{(\alpha)},\dots,A^n_{(\alpha)}\}$ *be the family of fuzzy clusters for some* $\alpha\in(0,1]$. *The point* $\tau^l_e \in A^l_\alpha$, *for which*

$$\tau^l_e = \arg\max_{x_i} \mu_{li}, \ \forall x_i \in A^l_\alpha \qquad (2.2)$$

is called a typical point of the fuzzy cluster $A^l_{(\alpha)}$, $\alpha\in(0,1]$, $l\in\{1,\dots,n\}$.

Obviously, a typical point of a fuzzy cluster does not depend on the value of the tolerance threshold. Moreover, a fuzzy cluster can have several typical points. Thus, the symbol e is an index of the typical point. The set $K(A^l_{(\alpha)}) = \{\tau^l_1,\dots,\tau^l_{|l|}\}$ of typical points of the fuzzy cluster $A^l_{(\alpha)}$ is a kernel of the fuzzy cluster and $card\left(K(A^l_{(\alpha)})\right)=|l|$ is the cardinality of the kernel. Obviously, if the fuzzy cluster have an unique typical point, then $|l|=1$.

Definition 2.3. *Let* $R^\alpha_{c(z)}(X) = \{A^l_{(\alpha)} \mid l=\overline{1,c},\, 2\le c\le n,\, \alpha\in(0,1]\}$ *be a family of* c *fuzzy clusters for some value of the tolerance threshold* α, $\alpha\in(0,1]$, *which are generated by some fuzzy tolerance relation* T *on the initial set of elements* $X=\{x_1,\dots,x_n\}$. *If the condition*

$$\sum_{l=1}^{c} \mu_{li} > 0, \ \forall x_i \in X \qquad (2.3)$$

is met for all fuzzy clusters $A^l_{(\alpha)} \in R^\alpha_z(X)$, $l=\overline{1,c}$, $c\le n$, *then this family is an allotment of the elements of the set* $X=\{x_1,\dots,x_n\}$ *among fuzzy clusters* $\{A^l_{(\alpha)},\, l=\overline{1,c},\, 2\le c\le n\}$ *for some value of the tolerance threshold* α.

It should be noted that the number of fuzzy clusters c in the allotment $R^\alpha_{c(z)}(X)$ can be fixed or unknown. Moreover, several allotments $R^\alpha_{c(z)}(X)$ can exist for some tolerance threshold α. That is why he symbol z is the index of an allotment.

The condition (2.3) requires that every object x_i, $i=1,\dots,n$, must be assigned to at least one fuzzy cluster $A^l_{(\alpha)}$, $l=\overline{1,c}$, $c\le n$, with the membership degree higher than zero. The condition $2\le c\le n$ requires that the number of fuzzy clusters in each allotment $R^\alpha_{c(z)}(X)$ must be at least two. Otherwise, the unique fuzzy cluster will contain all objects, possibly with different positive membership degrees.

The concept of an allotment is the central point of the method proposed though the concept introduced next should also be paid attention to.

Definition 2.4. *Allotment* $R_I^{\alpha}(X) = \{A_{(\alpha)}^l \mid l = \overline{1,n}, \alpha \in (0,1]\}$ *of the set of objects among* n *fuzzy clusters for some tolerance threshold* $\alpha \in (0,1]$ *is the initial allotment of the set* $X = \{x_1, \ldots, x_n\}$.

In other words, if the initial data are represented by some fuzzy tolerance matrix T, then the lines or columns of the matrix are fuzzy sets $A^l \subseteq X$, $l = \overline{1,n}$ and the level fuzzy sets $A_{(\alpha)}^l$, $l = \overline{1,c}$, $\alpha \in (0,1]$ are fuzzy clusters. These fuzzy clusters constitute an initial allotment for some tolerance threshold α and they can be considered as clustering components.

The definition of the allotment among the fuzzy clusters (2.3) is similar to the definition of the possibilistic partition (1.117). This fact was shown in [123]. On the other hand, let $A_{(\alpha)}^l \in R_{c(z)}^{\alpha}(X)$ be a fuzzy cluster with μ_{li} being its membership function in the sense of (2.1). So, $\exists \Pi$, $\forall x_i \in X$, $\mu_{li} = \Pi_l(\{x_i\}) = \pi_l(x_i)$. That is why the membership function of a fuzzy cluster can be interpreted as a possibility distribution.

In other words, the expression (2.1) defines a possibility distribution function for some $A_{(\alpha)}^l$, $l \in \{1, \ldots, c\}$, $\alpha \in (0,1]$ over the domain of discourse consisting of all objects $x_i \in X$. This possibility distribution will be denoted by $\pi_l(x_i)$ and its corresponding measure of possibility will be denoted by $\Pi_l(x_i)$, cf. [129]. That is why the allotment among the fuzzy clusters can be considered as a possibilistic partition, and the fuzzy clusters in the sense of Definition 2.1 are elements of a possibilistic partition. This fact was considered in [129] in detail. The concept of allotment will be used in our further considerations.

Thus, the problem of fuzzy cluster analysis can be defined in general as the problem of discovering the unique allotment $R_c^*(X)$, resulting from the classification process which corresponds to either a most natural allocation of objects among fuzzy clusters or to the analyst's opinion about the very essence and purpose of classification. In the first case, the number of fuzzy clusters c is not fixed. In the second case, the opinion determines the kind of the allotment sought and the number of fuzzy clusters c can be fixed.

If some allotment $R_{c(z)}^{\alpha}(X) = \{A_{(\alpha)}^l \mid l = \overline{1,c}, c \le n, \alpha \in (0,1]\}$ corresponds to the formulation of a specific problem, then this allotment is an adequate allotment. In particular, if the condition

$$\bigcup_{l=1}^{c} A_{\alpha}^l = X , \tag{2.4}$$

and the condition

$$card(A_{\alpha}^{l} \cap A_{\alpha}^{m}) = 0\,,\ \forall A_{(\alpha)}^{l}, A_{(\alpha)}^{m}\,,\ l \neq m\,,\ \alpha \in (0,1]\,, \tag{2.5}$$

are met for all fuzzy clusters $A_{(\alpha)}^{l}$, $l = \overline{1,c}$, of some allotment $R_{c(z)}^{\alpha}(X) = \{A_{(\alpha)}^{l} \mid l = \overline{1,c}, c \leq n\}$ for some value $\alpha \in (0,1]$, then this is the allotment among fully separate fuzzy clusters.

The fuzzy clusters in the sense of Definition 2.1 can have an intersection area. This fact was demonstrated in [116]. If the intersection area of any pair of different fuzzy clusters is an empty set, then the conditions (2.4) and (2.5) are met and the fuzzy clusters are called fully separate fuzzy clusters. Otherwise, the fuzzy clusters are called particularly separate fuzzy clusters and $w \in \{0,\ldots,n\}$ is the maximum number of elements in the intersection area of different fuzzy clusters. For $w = 0$ the fuzzy clusters are fully separate fuzzy clusters. Thus, the conditions (2.4) and (2.5) can be generalized for the case of particularly separate fuzzy clusters.

The condition

$$\sum_{l=1}^{c} card(A_{\alpha}^{l}) \geq card(X)\,,\ \forall A_{(\alpha)}^{l} \in R_{c(z)}^{\alpha}(X)\,,\ \alpha \in (0,1]\,,\ card(R_{c(z)}^{\alpha}(X)) = c\,, \tag{2.6}$$

and the condition

$$card(A_{\alpha}^{l} \cap A_{\alpha}^{m}) \leq w\,,\ \forall A_{(\alpha)}^{l}, A_{(\alpha)}^{m}\,,\ l \neq m\,,\ \alpha \in (0,1] \tag{2.7}$$

are therefore generalizations of the conditions (2.4) and (2.5). Obviously, if $w = 0$ in the conditions (2.6) and (2.7), then the conditions (2.4) and (2.5) are met.

The adequate allotment $R_{c(z)}^{\alpha}(X)$ for some value of the tolerance threshold $\alpha \in (0,1]$ is a family of fuzzy clusters which are elements of the initial allotment $R_{I}^{\alpha}(X)$ for α and the family of fuzzy clusters should satisfy the conditions (2.6) and (2.7). The construction of the adequate allotments $R_{c(z)}^{\alpha}(X) = \{A_{(\alpha)}^{l} \mid l = \overline{1,c}, c \leq n\}$, for each α, is a trivial problem of combinatorics.

The problem of intersection of pairs of different fuzzy clusters is very interesting from the theoretical point of view, and the following lemma was formulated and proved in [133].

Lemma 2.1. *Let $R_{c(z)}^{\alpha}(X) = \{A_{(\alpha)}^{1}, \ldots, A_{(\alpha)}^{c}\}$ be an allotment among fuzzy clusters for some $\alpha \in (0,1]$. If $A_{(\alpha)}^{l}, A_{(\alpha)}^{m} \in R_{c(z)}^{\alpha}(X)$, $l \neq m$, $\forall \alpha \in (0,1]$ are different fuzzy clusters with their corresponding kernels $K(A_{(\alpha)}^{l})$, $K(A_{(\alpha)}^{m})$, then the condition $K(A_{(\alpha)}^{l}) \cap K(A_{(\alpha)}^{m}) = \varnothing$ is met.*

Proof. Let $K(A^l_{(\alpha)}) = \{\tau^l_1, \ldots, \tau^l_{|l|}\}$ be a kernel of a fuzzy cluster $A^l_{(\alpha)} \in R^\alpha_{c(z)}(X)$ and $K(A^m_{(\alpha)}) = \{\tau^m_1, \ldots, \tau^m_{|l|}\}$ be a kernel of a fuzzy cluster $A^m_{(\alpha)} \in R^\alpha_{c(z)}(X)$. Let us assume that the condition $h(A^l_{(\alpha)}) \geq h(A^m_{(\alpha)})$ is met without loss of generality. That is why the conditions $\forall \tau^l_e \in K(A^l_{(\alpha)})$, $\mu_{A^l_{(\alpha)}}(\tau^l_e) = h(A^l_{(\alpha)})$ and $\forall \tau^m_{e'} \in K(A^m_{(\alpha)})$, $\mu_{A^m_{(\alpha)}}(\tau^m_{e'}) = h(A^m_{(\alpha)})$ are met.

Let assume that the condition $\tau^m_{e'} \in K(A^l_{(\alpha)})$ is met for some typical point $\tau^m_{e'}$. So, the condition $\mu_{A^l_{(\alpha)}}(\tau^m_{e'}) = \min\limits_{\tau^l_e \in K(A^l_{(\alpha)})} \mu_T(\tau^m_{e'}, \tau^l_e)$ will be met. Therefore, the condition $\forall x_i \in Supp(A^m_{(\alpha)}) \setminus K(A^m_{(\alpha)})$, $\mu_{A^m_{(\alpha)}}(x_i) < h(A^m_{(\alpha)})$, follows from Definition 2.2. We obtain $\forall x_i \in Supp(A^m_{(\alpha)})$, $\mu_{A^l_{(\alpha)}}(x_i) < \min\limits_{\tau^l_e \in K(A^l_{(\alpha)})} \mu_{A^l_{(\alpha)}}(\tau^l_e)$, leading to $Supp(A^m_{(\alpha)}) \subseteq Supp(A^l_{(\alpha)})$ and therefore $A^m_{(\alpha)} \subseteq A^l_{(\alpha)}$. In other words, the fuzzy cluster $A^m_{(\alpha)}$ should be included in the fuzzy cluster $A^l_{(\alpha)}$. Thus, the assumption that $\tau^m_{e'} \in K(A^l_{(\alpha)})$ contradicts the thesis of the lemma because the condition $l \neq m$ is met for fuzzy clusters $A^l_{(\alpha)}, A^m_{(\alpha)} \in R^\alpha_{c(z)}(X)$. Therefore, the lemma is proved. □

Several adequate allotments can exist. Thus, the problem consists in the selection of the unique adequate allotment $R^*_c(X)$ from the set B of adequate allotments, $B = \{R^\alpha_{c(z)}(X)\}$, which is the class of possible solutions of the concrete classification problem and $B = \{R^\alpha_{c(z)}(X)\}$ depends on the parameters of the classification problem. The selection of the unique adequate allotment $R^*_c(X)$ from the set $B = \{R^\alpha_{c(z)}(X)\}$ of adequate allotments must be made on the basis of an evaluation of allotments. The criterion which can be used for evaluation of allotments [114] is

$$F_1(R^\alpha_{c(z)}(X), \alpha) = \sum_{l=1}^{c} \frac{1}{n_l} \sum_{i=1}^{n_l} \mu_{li} - \alpha \cdot c, \tag{2.8}$$

where c is the number of fuzzy clusters in the allotment $R^\alpha_{c(z)}(X)$ and $n_l = card(A^l_\alpha)$, $A^l_{(\alpha)} \in R^\alpha_{c(z)}(X)$ is the number of elements in the support of the fuzzy cluster $A^l_{(\alpha)}$. The criterion

$$F_2(R^{\alpha}_{c(z)}(X),\alpha)=\sum_{l=1}^{c}\sum_{i=1}^{n_l}(\mu_{li}-\alpha)\,, \tag{2.9}$$

can also be used for the evaluation of allotments [123]. Both the criteria were proposed in [112] and those criteria are fuzzy generalizations of the criteria which have been proposed and investigated by Mirkin [78]. The maximum of the criterion (2.8) or criterion (2.9) corresponds to the best allotment of objects among c fuzzy clusters.

Therefore, the classification problem can formally be characterized as the determination of the solution $R^*_c(X)$ satisfying

$$R^*_c(X)=\arg\max_{R^{\alpha}_{c(z)}(X)\in B} F(R^{\alpha}_{c(z)}(X),\alpha)\,, \tag{2.10}$$

where $B=\{R^{\alpha}_{c(z)}(X)\}$ is the set of adequate allotments corresponding to the formulation of a specified classification problem and the criteria (2.8) and (2.9) are denoted by $F(R^{\alpha}_{c(z)}(X),\alpha)$.

The criterion (2.8) can be considered as the average total membership of objects in the fuzzy clusters of the allotment $R^{\alpha}_{c(z)}(X)$ minus $\alpha\cdot c$. The quantity $\alpha\cdot c$ regularizes with respect to the number of clusters c in the allotment $R^{\alpha}_{c(z)}(X)$. The criterion (2.9) can be considered as the total membership of objects in fuzzy clusters of the allotment $R^{\alpha}_{c(z)}(X)$ with an appreciation through the value α of a tolerance threshold. The condition (2.10) must be met for some unique allotment $R^{\alpha}_{c(z)}(X)\in B$. Otherwise, the number c of fuzzy clusters in the allotment sought $R^*_c(X)$ is not appropriate. This important condition was formulated in [115].

2.1.2 Illustrative Examples

The introduced definitions should be explained by a simple example [114]. Let $X=\{x_1,\ldots,x_5\}$ be the object set and the initial data matrix be as presented in Table 2.1. These data have originally appeared in [106]. The matrix given below represents a fuzzy tolerance T with the membership function $\mu_T(x_i,x_j)$.

Table 2.1 Matrix of the initial data

T	x_1	x_2	x_3	x_4	x_5
x_1	1.0				
x_2	0.8	1.0			
x_3	0.0	0.4	1.0		
x_4	0.1	0.0	0.0	1.0	
x_5	0.2	0.9	0.0	0.5	1.0

Thus, the columns or lines of the matrix of fuzzy tolerance are fuzzy sets on the finite universe X :

$$A^1 = \{(x_1,1.0),(x_2,0.8),(x_3,0.0),(x_4,0.1),(x_5,0.2)\},$$
$$A^2 = \{(x_1,0.8),(x_2,1.0),(x_3,0.4),(x_4,0.0),(x_5,0.9)\},$$
$$A^3 = \{(x_1,0.0),(x_2,0.4),(x_3,1.0),(x_4,0.0),(x_5,0.0)\},$$
$$A^4 = \{(x_1,0.1),(x_2,0.0),(x_3,0.0),(x_4,1.0),(x_5,0.5)\},$$
$$A^5 = \{(x_1,0.2),(x_2,0.9),(x_3,0.0),(x_4,0.5),(x_5,1.0)\},$$

where the zero values for the memberships of the fuzzy sets can be omitted. The α -level fuzzy sets are therefore

$$A^1_{(0.1)} = \{(x_1,1.0),(x_2,0.8),(x_4,0.1),(x_5,0.2)\},$$
$$A^2_{(0.1)} = \{(x_1,0.8),(x_2,1.0),(x_3,0.4),(x_5,0.9)\},$$
$$A^3_{(0.1)} = \{(x_2,0.4),(x_3,1.0)\},$$
$$A^4_{(0.1)} = \{(x_1,0.1),(x_4,1.0),(x_5,0.5)\},$$
$$A^5_{(0.1)} = \{(x_1,0.2),(x_2,0.9),(x_4,0.5),(x_5,1.0)\}$$

and are fuzzy clusters in the sense of Definition 2.1 for the value of the tolerance threshold $\alpha = 0.1$.

These fuzzy clusters constitue the initial allotment $R_I^{0.1}(X) = \{A^1_{(0.1)}, A^2_{(0.1)}, A^3_{(0.1)}, A^4_{(0.1)}, A^5_{(0.1)}\}$ for the tolerance threshold $\alpha = 0.1$. An allotment $R^{0.1}_{c(z)}(X)$ for the value of tolerance threshold $\alpha = 0.1$ is any family of fuzzy clusters which are elements of the initial allotment $R_I^{0.1}(X)$ and this family of fuzzy clusters should satisfy the conditions (2.6) and (2.7) for some value of $w \geq 0$. For example, the family of fuzzy clusters $R^{\alpha=0.1}_{c=2(z=1)}(X) = \{A^3_{(0.1)}, A^4_{(0.1)}\}$ can be considered as the allotment of the object set $X = \{x_1, \ldots, x_5\}$ among two fully separate fuzzy clusters. The object x_3 is the unique typical point τ^3 of the fuzzy cluster $A^3_{(0.1)}$ and the object x_4 is the unique typical point τ^4 of the fuzzy cluster $A^4_{(0.1)}$.

Let μ_{3i} be the membership function of the fuzzy cluster $A^3_{(0.1)}$ and μ_{4i} be the membership function of the fuzzy cluster $A^4_{(0.1)}$. Thus, the membership functions can be interpreted as the possibility distributions $\pi_l(x_i)$, $l \in \{3,4\}$, $\forall x_i \in X$. If $A^3_{0.1} = Supp(A^3_{(0.1)}) = \{x_2, x_3\}$ is the crisp subset of $X = \{x_1, \ldots, x_5\}$, then $\overline{A}^3_{0.1} = \{x_1, x_4, x_5\} = Supp(A^4_{(0.1)}) = A^4_{0.1}$. In other words, $A^3_{0.1} \cup A^4_{0.1} = X$ and $A^3_{0.1} \cap A^4_{0.1} = \varnothing$. So, the event $x_i \notin A^3_{0.1}$ is the contradiction of the event $x_i \in A^3_{0.1}$, $\forall x_i \in X$. That is why the event $x_i \notin A^3_{0.1}$ is equal to the event $x_i \in \overline{A}^3_{0.1}$. If the possibility distribution $\pi_3(x_i)$ is defined on the universe X, then the possibility degree of the event $x_2 \in A^3_{0.1}$ is

$$\Pi(x_2 \in A^3_{0.1}) = \sup_{x_i \in A^3_{0.1}} \pi_3(x_i) = \sup\{0.4, 1.0\} = 1.0,$$

whereas the necessity degree of the same event, $x_2 \in A^3_{0.1}$, is

$$\mathrm{N}(x_2 \in A^3_{0.1}) = \inf_{x_i \notin A^3_{0.1}} (1 - \pi_3(x_i)) = \inf\{1.0, 1.0, 1.0\} = 1.0.$$

The computation of the necessity degree $\mathrm{N}(x_2 \in A^3_{0.1})$ using the formula $\mathrm{N}(x_2 \in A^3_{0.1}) = 1 - \Pi(x_2 \in \overline{A}^3_{0.1})$ is the same and, in fact, $\mathrm{N}(x_2 \in A^3_{0.1}) = 1 - \Pi(x_2 \in \overline{A}^3_{0.1}) = 1 - \sup_{x_i \in \overline{A}^3_{0.1}} \pi_3(x_i) = 1 - \sup\{0.0, 0.0, 0.0\} = 1.0$.

By computing $\Pi(x_2 \in \overline{A}^3_{0.1})$ and $\mathrm{N}(x_2 \in \overline{A}^3_{0.1})$, we obtain that $\Pi(x_2 \in \overline{A}^3_{0.1}) = \sup_{x_i \in \overline{A}^3_{0.1}} \pi_3(x_i) = 0.0$ and $\mathrm{N}(x_2 \in \overline{A}^3_{0.1}) = 1 - \Pi(x_2 \in A^3_{0.1}) = 0.0$.

So, the equations $\Pi(A) + \Pi(\overline{A}) \geq 1$ and $\mathrm{N}(A) + \mathrm{N}(\overline{A}) \leq 1$ [33] are met for the event $x_2 \in A^3_{0.1}$ and its contradiction $x_2 \in \overline{A}^3_{0.1}$.

On the other hand, if we assume that $w = 1$ in the condition (2.7), then the allotments $R^{\alpha=0.1}_{c=2(z=1)}(X) = \{A^1_{(0.1)}, A^3_{(0.1)}\}$ and $R^{\alpha=0.1}_{c=2(z=2)}(X) = \{A^3_{(0.1)}, A^5_{(0.1)}\}$ among two particularly separate fuzzy clusters can be considered as adequate allotments, too.

Let us consider the detection of the allotment among three fully separate fuzzy clusters. Any allotment among three fully separate fuzzy clusters will be the adequate allotment for the problem of classification. The adequate allotments exist for $\alpha = 0.5$ and $\alpha = 0.8$ and the allotments are presented in Table 2.2.

Table 2.2 Fragment of the classification process

Value of the tolerance threshold	The initial allotment, $R_I^{\alpha}(X)$	The adequate allotment, $R_{c(z)}^{\alpha}(X)$	Value of the criterion	
			(2.8)	(2.9)
0.5	$A_{(0.5)}^1 = \{(x_1,1.0),(x_2,0.8)\}$, $A_{(0.5)}^2 = \{(x_1,0.8),(x_2,1.0),(x_5,0.9)\}$, $A_{(0.5)}^3 = \{(x_3,1.0)\}$, $A_{(0.5)}^4 = \{(x_4,1.0),(x_5,0.5)\}$, $A_{(0.5)}^5 = \{(x_2,0.9),(x_4,0.5),(x_5,1$	$R_{3(1)}^{0.5}(X) = \{A_{(0.5)}^1, A_{(0.5)}^3, A_{(0.5)}^4\}$	1.15	1.80
0.8	$A_{(0.8)}^1 = \{(x_1,1.0),(x_2,0.8)\}$, $A_{(0.8)}^2 = \{(x_1,0.8),(x_2,1.0),(x_5,0.9)\}$, $A_{(0.8)}^3 = \{(x_3,1.0)\}$, $A_{(0.8)}^4 = \{(x_4,1.0)\}$, $A_{(0.8)}^5 = \{(x_2,0.9),(x_5,1.0)\}$	$R_{3(1)}^{0.8}(X) = \{A_{(0.8)}^2, A_{(0.8)}^3, A_{(0.8)}^4\}$	0.50	0.70

The number of fuzzy clusters c in the allotment sought is a parameter of the classification. That is why the set of adequate allotments will be denoted as $B(c) = \{R_{c(z)}^{\alpha}(X)\}$. The set $B(c=3)$ of adequate allotments includes therefore two adequate allotments, $B(c=3) = \{R_{3(1)}^{0.5}(X), R_{3(1)}^{0.8}(X)\}$. Values of the criteria (2.8) and (2.9) are maximal for the allotment $R_{3(1)}^{0.5}(X) = \{A_{(0.5)}^1, A_{(0.5)}^3, A_{(0.5)}^4\}$. So, the allotment $R_{3(1)}^{0.5}(X)$ should be selected as the result $R_c^*(X)$ of classification. The matrix of object assignments is presented in Table 2.3.

Table 2.3 Matrix of object assignments

Object	Membership degree		
	Class 1	Class 2	Class 3
x_1	1.0	0.0	0.0
x_2	0.8	0.0	0.0
x_3	0.0	1.0	0.0
x_4	0.0	0.0	1.0
x_5	0.0	0.0	0.5

The fuzzy cluster $A^1_{(0.5)} = \{(x_1,1.0),(x_2,0.8)\}$ corresponds to the first class and the object x_1 is the unique typical point τ^1 of the fuzzy cluster. The fuzzy cluster $A^3_{(0.5)} = \{(x_3,1.0)\}$ corresponds to the second class and the unique object which belongs to the class is its typical point, $x_3 = \tau^3$. The fuzzy cluster $A^4_{(0.5)} = \{(x_4,1.0),(x_5,0.5)\}$ corresponds to the third class and the object x_4 is the unique typical point τ^4 of the class. These results can be presented as in Figure 2.1 in which the membership functions of three fuzzy clusters of the detected allotment $R^*_c(X)$ are shown.

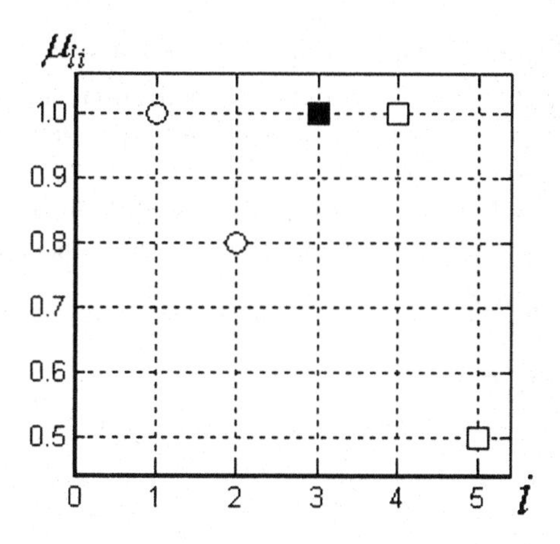

Fig. 2.1 Diagram of object assignments

The membership values of the first class are represented by ○, the membership values of the second class are represented by ■ and the membership values of the third class are represented by □.

2.1.3 A Note on the Data Preprocessing

The matrix of fuzzy tolerance $T = [\mu_T(x_i, x_j)]$, $i, j = 1,\ldots,n$ is the matrix of initial data for the clustering method proposed. However, the data can be presented as the matrix of attributes $\hat{X}_{n \times m_1} = [\hat{x}_i^{t_1}], i = 1,\ldots,n$, $t_1 = 1,\ldots,m_1$, in which the value $\hat{x}_i^{t_1}$ is the value of the t_1-th attribute for the i-th object. Thus, the proposed approach to clustering can be used with the two-way data (1.69), by choosing a suitable metric to measure the similarity.

On the one hand, the two-way data can be normalized as follows:

$$x_i^{t_1} = \frac{\hat{x}_i^{t_1}}{\max\limits_i \hat{x}_i^{t_1}}. \tag{2.11}$$

while the two-way data can be normalized using the formula

$$x_i^{t_1} = \frac{\hat{x}_i^{t_1} - \min\limits_i \hat{x}_i^{t_1}}{\max\limits_i \hat{x}_i^{t_1} - \min\limits_i \hat{x}_i^{t_1}}. \tag{2.12}$$

The data normalization method due to (2.11) is appropriate in the case of non-negative values $\hat{x}_i^{t_1}$ in the two-way data matrix. Thus, each object can be considered as a fuzzy set x_i, $i = 1,\ldots,n$ and $x_i^{t_1} = \mu_{x_i}(x^{t_1}) \in [0,1]$, $i = 1,\ldots,n$, $t_1 = 1,\ldots,m_1$ are the membership functions of these fuzzy sets. Of course, some other methods for the two-way data normalization are described in the bibliographical sources. Different methods for the data normalization are considered, for example, by Walesiak [151].

The matrix of coefficients of the pair wise dissimilarity between objects $I = [\mu_I(x_i, x_j)]$, $i, j = 1,\ldots,n$ can be obtained after the application of some distance function to the matrix of normalized data $X_{n \times m_1} = [\mu_{x_i}(x^{t_1})]$, $i = 1,\ldots,n$, $t_1 = 1,\ldots,m_1$.

The most widely used distances for the fuzzy sets x_i, x_j, $i, j = 1,\ldots,n$ in $X = \{x_1,\ldots,x_n\}$ are:

- The normalized Hamming distance:

$$l(x_i, x_j) = \frac{1}{m_1} \sum_{t_1=1}^{m_1} |\mu_{x_i}(x^{t_1}) - \mu_{x_j}(x^{t_1})|, i, j = 1,\ldots,n, \tag{2.13}$$

- The normalized Euclidean distance:

$$e(x_i, x_j) = \sqrt{\frac{1}{m_1} \sum_{t_1=1}^{m_1} \left(\mu_{x_i}(x^{t_1}) - \mu_{x_j}(x^{t_1}) \right)^2}, i, j = 1, \ldots, n, \quad (2.14)$$

- The squared normalized Euclidean distance:

$$\varepsilon(x_i, x_j) = \frac{1}{m_1} \sum_{t_1=1}^{m_1} \left(\mu_{x_i}(x^{t_1}) - \mu_{x_j}(x^{t_1}) \right)^2, \; i, j = 1, \ldots, n. \quad (2.15)$$

These distances have been considered by Kacprzyk [58] in detail. It should be noted that the normalized Hamming distance (2.13) and the normalized Euclidean distance (2.14) satisfy the following three conditions:

- Non-negativity:

$$d(x_i, x_j) \geq 0, \forall x_i, x_j \in X, \; x_i \neq x_j, \quad (2.16)$$

- Symmetry:

$$d(x_i, x_j) = d(x_j, x_i), \forall x_i, x_j \in X, \quad (2.17)$$

- Transitivity:

$$d(x_i, x_k) \leq d(x_i, x_j) + d(x_j, x_k), \forall x_i, x_j, x_k \in X, \quad (2.18)$$

- Antireflexivity:

$$d(x_i, x_i) = 0, \forall x_i \in X, \quad (2.19)$$

where $d(x_i, x_j)$ denotes a distance function between two elements x_i and x_j of the set X. On the other hand, the squared normalized Euclidean distance (2.15) does not satisfy the transitivity condition (2.18). This fact was demonstrated by Kaufmann [60].

If the initial data is presented by the relation matrix (1.70), then the relational data can be normalized as follows:

$$\rho_{ij} = \frac{\left(\rho_{ij} - \min_{i,j} \rho_{ij} \right)}{\max_{i,j} \rho_{ij} - \min_{i,j} \rho_{ij}}, \quad (2.20)$$

where the general notation ρ_{ij} is used for the designation of pair-wise dissimilarities or the similarity coefficients. If $\rho_{ii} = 0, \; \forall i$, then the relation matrix $\rho_{n\times n} = [\rho_{ij}]$ is the matrix of fuzzy intolerance $I = [\mu_I(x_i, x_j)]$, $i, j = 1,\ldots,n$. The matrix of fuzzy tolerances $T = [\mu_T(x_i, x_j)]$, $i, j = 1,\ldots,n$ can be obtained after the application of the complement operation (1.26) to the matrix of fuzzy intolerances $I = [\mu_I(x_i, x_j)]$, $i, j = 1,\ldots,n$.

2.2 Relational Clustering Algorithms

Three relational clustering algorithms are described in the corresponding three subsections of the present section. Measures for the evaluation of fuzzy clusters are considered in the fourth subsection and cluster validity indices are introduced in the fifth subsection. The sixth subsection includes the description of some well-known data sets and results of their processing by the relational heuristic algorithms of possibilistic clustering in comparison with some fuzzy clustering methods and objective function-based algorithms of possibilistic clustering.

2.2.1 *The Basic Algorithm*

The detection of a fixed c number of fuzzy clusters can be considered as the purpose of classification. An adequate allotment $R^{\alpha}_{c(z)}(X)$ is any allotment among c fuzzy clusters and the result of classification will be denoted as $R^{*}_{c}(X)$ in the case considered. The basic version of direct clustering algorithm was described in [114] where the clustering procedure was called the AFC-algorithm. Note at this point that the name of AFC-algorithm was used for the fuzzy clustering algorithm which was proposed by Davé [29]. That is why the direct algorithm for detecting the allotment among c fuzzy clusters has been called the D-AFC(c)-algorithm in further publications. The basic version of the clustering algorithm is a sevenstep procedure of classification.

The D-AFC(c)-Algorithm

1. Calculate α-level values of the fuzzy tolerance T and construct the sequence $0 < \alpha_0 < \alpha_1 < \ldots < \alpha_\ell < \ldots < \alpha_Z \le 1$ of α-levels;
2. Construct an initial allotment $R^{\alpha}_{I}(X) = \{A^{l}_{(\alpha)} \mid l = \overline{1,n}\}$, $\alpha = \alpha_\ell$, for each value α_ℓ from the sequence $0 < \alpha_0 < \alpha_1 < \ldots < \alpha_\ell < \ldots < \alpha_Z \le 1$;
3. Let $w := 0$;
4. Construct allotments $R^{\alpha}_{c(z)}(X) = \{A^{l}_{(\alpha)} \mid l = \overline{1,c}, c \le n\}$, $\alpha = \alpha_\ell$, which satisfy the conditions (2.6) and (2.7) for each value α_ℓ from the sequence $0 < \alpha_0 < \alpha_1 < \ldots < \alpha_\ell < \ldots < \alpha_Z \le 1$;

5. Construct the class of possible solutions of the classification problem $B(c) = \{R^{\alpha}_{c(z)}(X)\}$, $\alpha \in \{\alpha_1, \ldots, \alpha_Z\}$, for the given number of fuzzy clusters c and different values of the tolerance threshold, $\alpha \in \{\alpha_1, \ldots, \alpha_Z\}$, as follows:

   if for some allotment $R^{\alpha}_{c(z)}(X)$ the condition $card(R^{\alpha}_{c(z)}(X)) = c$ is met

   then $R^{\alpha}_{c(z)}(X) \in B(c)$

   else let $w := w + 1$ and go to Step 4;
6. Calculate the value of some criterion $F(R^{\alpha}_{c(z)}(X), \alpha)$ for each allotment $R^{\alpha}_{c(z)}(X) \in B(c)$;
7. The result $R^*_c(X)$ of classification is formed as follows:

   if for some unique allotment $R^{\alpha}_{c(z)}(X) \in B(c)$ the condition (2.10) is met

   then the allotment is a result of classification $R^*_c(X)$
   else the number c of classes is suboptimal.

The ordered sequence $0 < \alpha_0 < \ldots < \alpha_\ell < \ldots < \alpha_Z \leq 1$ of α-level values can be constructed from the values of the membership function $\mu_T(x_i, x_j)$ using the formulae (1.56) – (1.58) in the first step of the D-AFC(c)-algorithm. The allotment $R^*_c(X)$ among c partially separate fuzzy clusters and the value of the tolerance threshold $\alpha \in (0,1]$ are results of classification obtained from the D-AFC(c)-algorithm.

2.2.2 An Algorithm with Partial Supervision

The partially supervised fuzzy clustering plays an unique role in discovering the structure in data realized in the presence of labeled patterns. This may be very useful, for example, in speech recognition systems. Some other problems related to robotics and automation can also be successfully solved by using partially supervised clustering methods.

The idea of partial supervision in fuzzy clustering was originated by Pedrycz [88] and later developed by different researchers. For example, an original semi-supervised modification of the FCM-algorithm was proposed by Bensaid, Hall, Bezdek and Clarke [9]. The method is well suited to problems such as image segmentation. In particular, the procedure was effectively and efficiently applied to the segmentation of magnetic resonance images [9]. On the other hand, a semi-supervised modification of the locality-weighted fuzzy c-means have been proposed by Huang and Zhang [51] who have described the corresponding SLFCM-algorithm and the SLFCM-S-algorithm. Very interesting and important results in the area of fuzzy clustering with partial supervision have been presented by Bouchachia and Pedrycz [12].

The modification of the D-AFC(c)-algorithm in the presence of labeled objectshas been proposed in [123]. A direct algorithm for detecting the allotment among c fuzzy clusters with partial supervisionis called the D-AFC-PS(c)-algorithm. Therefore, a mechanism of partial supervision for the method should be considered in detail.

Let us consider a subset of labeled objects $X_L = \{x_{L(1)}, \ldots, x_{L(c)}\}$ and $X_L \subset X$. The condition $card(X_L) = c$ must be met for this subset. Let the membership degrees $y_{l(j)}$, $l = 1, \ldots, c$, $j = 1, \ldots, c$, correspond to each labeled object $x_{L(j)} \in X_L$, $j = 1, \ldots, c$, as follows: if $x_i \in X_L$ and $x_i = x_{L(j)}$, then the values of $y_{l(j)}$ are given by researcher. So, the detection of a fixed number c of fuzzy clusters can be considered as the aim of classification and each labeled object must be assigned to a unique fuzzy cluster. Moreover, for each labeled object $x_i = x_{L(j)}$ its membership degree μ_{li}, $l = 1, \ldots, c$, $i = 1, \ldots, n$, in the allotment sought $R_c^*(X)$ must be greater than an a priori determined membership grade $y_{l(j)} \in (0,1]$. The D-AFC-PS(c)-algorithm is a seven-step classification procedure as given below.

The D-AFC-PS(c)-Algorithm

1. Calculate α-level values of the fuzzy tolerance T and construct the sequence $0 < \alpha_0 < \alpha_1 < \ldots < \alpha_\ell < \ldots < \alpha_Z \leq 1$ of α-levels;
2. Construct an initial allotment $R_I^\alpha(X) = \{A_{(\alpha)}^l \mid l = \overline{1,n}\}$, $\alpha = \alpha_\ell$, for each value α_ℓ from the sequence $0 < \alpha_0 < \alpha_1 < \ldots < \alpha_\ell < \ldots < \alpha_Z \leq 1$;
3. Let $w := 0$;
4. Construct allotments $R_{c(z)}^\alpha(X) = \{A_{(\alpha)}^l \mid l = \overline{1,c}, c \leq n\}$, $\alpha = \alpha_\ell$, which satisfy the conditions (2.6) and (2.7) for each value α_ℓ from the sequence $0 < \alpha_0 < \alpha_1 < \ldots < \alpha_\ell < \ldots < \alpha_Z \leq 1$;
5. Construct the class of possible solutions of the classification problem $B(c) = \{R_{c(z)}^\alpha(X)\}$, $\alpha \in \{\alpha_1, \ldots, \alpha_Z\}$ for the given number of fuzzy clusters, c, and different values of the tolerance threshold, $\alpha \in \{\alpha_1, \ldots, \alpha_Z\}$, as follows:

   if for some allotment $R_{c(z)}^\alpha(X)$ the condition $card(R_{c(z)}^\alpha(X)) = c$ is met

   and for each labeled object $x_{L(j)} = x_i$, $j = \overline{1,c}$, $i \in \{1, \ldots, n\}$, the condition $\mu_{li} \geq y_{lj}$, $A_{(\alpha)}^l \in R_{c(z)}^\alpha(X)$, $l = 1, \ldots, c$, is met

   then $R_{c(z)}^\alpha(X) \in B(c)$

   else let $w := w + 1$ and go to Step 4;

6. Calculate the value of some criterion $F(R^{\alpha}_{c(z)}(X),\alpha)$ for each allotment $R^{\alpha}_{c(z)}(X) \in B(c)$;
7. The result $R^{*}_{c}(X)$ of classification is formed as follows:

   if for some unique allotment $R^{\alpha}_{c(z)}(X) \in B(c)$ the condition (2.10) is met

   then the allotment is a result of classification $R^{*}_{c}(X)$
   else the number c of classes is suboptimal.

Obviously, the D-AFC-PS(c)-algorithm does not differ significantly from the basic version of the clustering procedure. Different techniques for constructing a subset of labeled objects $X_L \subset X$ are considered, for example, in [127], [138] and [28].

2.2.3 An Algorithm for Detecting the Minimal Number of Disjointed Clusters

The detection of the unknown minimal number of compact and well-separated fuzzy clusters can be considered as the aim of classification in some situations. The following concept was introduced in [126].

Definition 2.5. *Allotment* $R^{\alpha}_{P}(X) = \{A^{l}_{(\alpha)} \mid l = \overline{1,c}\}$ *of the set of objects among the minimal number* c, $2 \leq c \leq n$, *of fully separate fuzzy clusters for some tolerance threshold* $\alpha \in (0,1]$ *is the principal allotment of the set* $X = \{x_1,\ldots,x_n\}$

Thus, the corresponding direct algorithm for detecting the principal allotment among fuzzy clusters was introduced in [135], called the D-PAFC-algorithm. It is a five-step classification procedure as given below:

The D-PAFC-Algorithm

1. Calculate α-level values of the fuzzy tolerance T and construct the sequence $0 < \alpha_0 < \alpha_1 < \ldots < \alpha_\ell < \ldots < \alpha_Z \leq 1$ of α-levels; let $\ell := 1$;
2. Construct the initial allotment $R^{\alpha}_{I}(X) = \{A^{l}_{(\alpha)} \mid l = \overline{1,n}\}$, $\alpha = \alpha_\ell$, for each value α_ℓ from the sequence $0 < \alpha_0 < \alpha_1 < \ldots < \alpha_\ell < \ldots < \alpha_Z \leq 1$;
3. The following condition is checked:

   if for some fuzzy cluster $A^{l}_{(\alpha)} \in R^{\alpha}_{I}(X)$, $\alpha = \alpha_\ell$, the condition $n_l = n$ is met

   then let $\ell := \ell + 1$ and go to Step 2
   else construct the allotments which satisfy the conditions (2.4) and (2.5);

4. The following condition is checked:

   if for α_ℓ the allotments $R^{\alpha}_{c(z)}(X)$ satisfying the conditions (2.4) and (2.5) are not constructed
   then let $\ell := \ell + 1$ and go to Step 2
   else construct the class of possible solutions of the classification problem $B^{\alpha_\ell} = \{R^{\alpha}_{c(z)}(X)\}$, $\alpha = \alpha_\ell$ for α_ℓ;

5. The following condition is checked:

   if condition $card(B^{\alpha_\ell}) > 1$ is met

   then calculate the value of some criterion $F(R^{\alpha}_{c(z)}(X), \alpha)$ for every allotment $R^{\alpha}_{c(z)}(X) \in B^{\alpha_\ell}$

   and the classification result $R^*_c(X)$ is constructed as follows:

   if for some unique allotment $R^{\alpha}_{c(z)}(X) \in B^{\alpha_\ell}$ the condition (2.10) is met
   then the allotment is a solution $R^*_c(X)$ of the classification problem
   else **if** condition $card(B^{\alpha_\ell}) = 1$ is met
   then the unique allotment $R^{\alpha}_{c(z)}(X) \in B^{\alpha_\ell}$ is a solution $R^*_c(X)$ of the classification problem.

The principal allotment $R^{\alpha}_{P}(X) = \{A^l_{(\alpha)} \mid l = \overline{1,c}\}$ among the unknown least number of fully separate fuzzy clusters and the value of the tolerance threshold $\alpha \in (0,1]$ are results of classification.

2.2.4 *Evaluation of the Fuzzy Clusters*

It is should be noted, that computational accuracy must be taken into account in data processing by clustering algorithms. Computational accuracy can be dealt with via the value of the accuracy threshold $\varepsilon \in (0,1]$. If we decrease the value ε, computational accuracy increases. The membership degrees μ_{li}, $l = 1,\ldots,c$, $i = 1,\ldots,n$, the value of tolerance threshold α, and the number of typical points $|l|$ in each fuzzy cluster $A^l_{(\alpha)} \in R^*_c(X)$ depend on the value of the accuracy threshold ε. An allotment $R^*_c(X)$ can be characterized by the value of tolerance threshold $\alpha \in (0,1]$ that is increasing with a decreasing accuracy, i.e., for $\varepsilon \to 1$ we have $\alpha \to 1$. On the other hand, a fuzzy cluster $A^l_{(\alpha)} \in R^*_c(X)$ can be characterized by a kernel $K(A^l_{(\alpha)})$ and the number of typical points of the fuzzy cluster is decreasing with an increasing accuracy, i.e., for $\varepsilon \to 0$ we have $card\left(K(A^l_{(\alpha)})\right) \to 1$. So, the accuracy threshold ε can be used as a parameter for clustering algorithms. This was demonstrated by Damaratski and Novikau [27]. Moreover, the accuracy threshold ε can be considered as an analog of the fuzziness index γ in fuzzy objective functions in optimization based clustering algorithms. This idea was outlined in [142].

The classification result is to be interpreted from a substantial point of view. Some formal criteria can be useful for this purpose. For example, the most appropriate distance between fuzzy sets for the data preprocessing can be selected on the basis of evaluation of the results of classification. A problem of the evaluation of fuzzy clusters was considered in [120].

The qualitative inspection of fuzzy clustering results can be done, e.g., with a linear index of fuzziness or a quadratic index of fuzziness, used for the evaluation of fuzziness degree of fuzzy clusters. These two indexes are considered by Kaufmann [60]. A modification of the linear index of fuzziness is defined in [120] as

$$I_L(A^l_{(\alpha)}) = \frac{2}{n_l} \cdot d_H(A^l_{(\alpha)}, \underline{A}^l_{(\alpha)}), \tag{2.21}$$

where $n_l = card(A^l_\alpha)$, $A^l_{(\alpha)} \in R^*_c(X)$, is the number of objects in the fuzzy cluster $A^l_{(\alpha)}$ and $d_H(A^l_{(\alpha)}, \underline{A}^l_{(\alpha)})$ is the Hamming distance

$$d_H(A^l_{(\alpha)}, \underline{A}^l_{(\alpha)}) = \sum_{x_i \in A^l_\alpha} \left| \mu_{li} - \mu_{\underline{A}^l_{(\alpha)}}(x_i) \right| \tag{2.22}$$

between the fuzzy cluster $A^l_{(\alpha)}$ and the crisp set $\underline{A}^l_{(\alpha)}$ that is the nearest to the fuzzy cluster $A^l_{(\alpha)}$. The membership function of the crisp set $\underline{A}^l_{(\alpha)}$ can be defined as

$$\mu_{\underline{A}^l_{(\alpha)}}(x_i) = \begin{cases} 0, & \mu_{A^l_{(\alpha)}}(x_i) \le 0.5 \\ 1, & \mu_{A^l_{(\alpha)}}(x_i) > 0.5 \end{cases}, \forall x_i \in A^l_\alpha, \tag{2.23}$$

where $\alpha \in (0,1]$.

The modified quadratic index of fuzziness is defined in [120] as

$$I_Q(A^l_{(\alpha)}) = \frac{2}{\sqrt{n_l}} \cdot d_E(A^l_{(\alpha)}, \underline{A}^l_{(\alpha)}), \tag{2.24}$$

where $n_l = card(A_\alpha^l)$, $A_{(\alpha)}^l \in R_c^*(X)$ and $d_E(A_{(\alpha)}^l, \underline{A}_{(\alpha)}^l)$ is the Euclidean distance

$$d_E(A_{(\alpha)}^l, \underline{A}_{(\alpha)}^l) = \sqrt{\sum_{x_i \in A_\alpha^l} \left(\mu_{li} - \mu_{\underline{A}_{(\alpha)}^l}(x_i) \right)^2} \quad (2.25)$$

between the fuzzy cluster $A_{(\alpha)}^l$ and the crisp set $\underline{A}_{(\alpha)}^l$ which is defined by the formula (2.23).

For each fuzzy cluster $A_{(\alpha)}^l$ in the allotment $R_c^*(X)$, evidently, the following conditions are met:

$$0 \le I_L(A_{(\alpha)}^l) \le 1, \quad (2.26)$$

$$0 \le I_Q(A_{(\alpha)}^l) \le 1. \quad (2.27)$$

The indexes (2.21) and (2.24) show the degree of fuzziness of fuzzy clusters which are elements of the allotment $R_c^*(X)$. Obviously, $I_L(A_{(\alpha)}^l) = I_Q(A_{(\alpha)}^l) = 0$ for a crisp set $A_{(\alpha)}^l \in R_c^*(X)$. Otherwise, if $\mu_{li} = 0.5$, $\forall x_i \in A_\alpha^l$, then the fuzzy cluster $A_{(\alpha)}^l \in R_c^*(X)$ is fuzzy set with the maximum value of the degree of fuzziness, and $I_L(A_{(\alpha)}^l) = I_Q(A_{(\alpha)}^l) = 1$.

The density of fuzzy cluster was defined in [120] as follows:

$$D(A_{(\alpha)}^l) = \frac{1}{n_l} \sum_{x_i \in A_\alpha^l} \mu_{li}, \quad (2.28)$$

where $n_l = card(A_\alpha^l)$, $A_{(\alpha)}^l \in R_c^*(X)$, and the membership degree μ_{li} is defined by the formula (2.1).

It is obvious that the condition

$$0 < D(A_{(\alpha)}^l) \le 1, \quad (2.29)$$

is met for each fuzzy cluster $A_{(\alpha)}^l$ in $R_c^*(X)$. Moreover, $D(A_{(\alpha)}^l) = 1$ for a crisp set $A_{(\alpha)}^l \in R_c^*(X)$ for any tolerance threshold α, $\alpha \in (0,1]$. The density of a fuzzy cluster shows an average membership degree of elements of a fuzzy cluster.

2.2.5 Validity Measures for the Basic Algorithm

The most “plausible” number c of fuzzy clusters in the allotment $R_c^*(X)$ sought can be considered as the cluster validity problem for the D-AFC(c)-algorithm. The number c of fuzzy clusters and their compactness are contradictory conditions in the classification of n objects. If compact classes are searched, the most appropriate solution can be obtained with n classes of one object. Obviously, such a solution is not useful. So, the number c of fuzzy clusters must be determined by considering conditions: first, the number of fuzzy clusters c in the allotment $R_c^*(X)$ sought must be as low as possible, and, second, the membership function of fuzzy clusters of some allotment among c fuzzy clusters must be sharper than the membership function of the fuzzy clusters of allotments for other numbers of fuzzy clusters.

Let $R_c^*(X)$ be the allotment which corresponds to the result of classification for the given number c of fuzzy clusters and R^c be the set of all allotments $R_c^*(X)$ among c, $c \in \{2,\ldots,n\}$, fuzzy clusters.

A cluster validity measure can be defined as a mapping $V : R^c \mapsto \Re$ which can be used to rank the validity of various allotments $R_c^*(X)$. Validity measures can be obtained from the indexes which are defined in the previous section.

The fuzziness of the allotment $R_c^*(X)$ among c fuzzy clusters can be evaluated as the sum of indexes of fuzziness of fuzzy clusters of the allotment $R_c^*(X)$. So, the linear measure of fuzziness of the allotment must be based on the formula (2.21), and the measure can be defined as follows:

$$V_{LMF}\left(R_c^*(X);c\right) = \sum_{A_{(\alpha)}^l \in R_c^*(X)} \left(I_L(A_{(\alpha)}^l)\right) = \sum_{A_{(\alpha)}^l \in R_c^*(X)} \left(\frac{2}{n_l} \cdot d_H(A_{(\alpha)}^l, \underline{A}_{(\alpha)}^l)\right). \quad (2.30)$$

The linear measure of fuzziness of the allotment, $V_{LMF}(R_c^*(X);c)$, was introduced in [136].

On the other hand, the quadratic measure of fuzziness of the allotment [144] can be defined on the analogy of the linear measure of fuzziness (2.30):

$$V_{QMF}\left(R_c^*(X);c\right) = \sum_{A_{(\alpha)}^l \in R_c^*(X)} \left(I_Q(A_{(\alpha)}^l)\right) = \sum_{A_{(\alpha)}^l \in R_c^*(X)} \left(\frac{2}{\sqrt{n_l}} \cdot d_E(A_{(\alpha)}^l, \underline{A}_{(\alpha)}^l)\right), \quad (2.31)$$

where $I_Q(A_{(\alpha)}^l)$ is the modified quadratic index of fuzziness (2.24).

The justification of both measures of fuzziness is intuitively clear. Obviously, the fuzziness of each fuzzy cluster $A_{(\alpha)}^l \in R_c^*(X)$ depends on the size of the fuzzy cluster. The number of objects n_l in each fuzzy cluster $A^l_{(\alpha)} \in R^*_c(X)$ is decreasing with the increase of the number c of fuzzy clusters in the allotment $R^*_c(X)$. That is why the fuzziness of each fuzzy cluster $A^l_{(\alpha)} \in R^*_c(X)$ is decreasing with the increase of the number c of fuzzy clusters. In other words, for $c \to n$ we have $n_l \to 1$ and $I_L(A^l_{(\alpha)}) \to 0$, $I_Q(A^l_{(\alpha)}) \to 0$ for all $A^l_{(\alpha)}$, $l \in \{1,\dots,c\}$. Therefore, for $c \to n$ we have $V_{LMF}\left(R^*_c(X);c\right) \to 0$ and $V_{QMF}\left(R^*_c(X);c\right) \to 0$, and the maximal value of a measure of fuzziness of the allotment $R^*_c(X)$ corresponds to the minimal number c of compact fuzzy clusters $A^l_{(\alpha)} \in R^*_c(X)$, $l = 1,\dots,c$, in the allotment sought. Using $V_{LMF}(R^*_c(X);c)$ or $V_{QMF}\left(R^*_c(X);c\right)$, the optimal number of fuzzy clusters can be obtained by maximizing the index.

The density of a fuzzy cluster (2.28) can be considered as the basis for a validity measure, cf. [144]. The validity measure must be taking into account the compactness of fuzzy clusters which is characterized by their density. The density of each fuzzy cluster $A^l_{(\alpha)} \in R^*_c(X)$ is increasing with the increase of the number c of fuzzy clusters. So, for $c \to n$ we have $D(A^l_{(\alpha)}) \to 1$ for all $A^l_{(\alpha)}$, $l \in \{1,\dots,c\}$ and for $c \to n$ we have $\alpha \to 1$. Thus, the value of the tolerance threshold α must be taken into account.

The validity measure can be defined as the ratio of the sum of densities of fuzzy clusters of some allotment to the number of fuzzy clusters minus the value of the tolerance threshold α. However, the case of particularly separate fuzzy clusters must be taken into account. That is why the sum of membership degrees of elements in the intersection areas of fuzzy clusters must be calculated and the ratio of the number c of fuzzy clusters in the allotment $R^*_c(X)$ to the number of elements of the data set must be taken into account, too.

Therefore, the measure of separation and compactness of the allotment can be defined in the following way:

$$V_{MSC}(R^*_c(X);c) = \frac{\sum\limits_{A^l_{(\alpha)} \in R^*_c(X)} D(A^l_{(\alpha)})}{c} + \frac{c}{n} \sum_{x_j \in \Theta} \mu_{lj} - \alpha\,, \tag{2.32}$$

where Θ is the set of elements x_j, $j \in \{1,\dots,n\}$, in all intersection areas of different fuzzy clusters.

The validity measure (2.32) has three components. The first component is the ratio of the sum of densities of fuzzy clusters of an allotment $R_c^*(X)$ to the number of fuzzy clusters c. The second is a penalty term which regularizes with respect to the membership values of elements in the intersection areas of fuzzy clusters. The third component is the value of the tolerance threshold. Note that the value of the second component of the measure of separation and compactness (2.32) will be equal zero in the case of fully separate fuzzy clusters in the allotment $R_{c(z)}^{\alpha}(X) \in B(c)$.

On the other hand, several allotments $R_{c(z)}^{\alpha}(X)$ among c fuzzy clusters cannot exist as solutions of the classification problem because the condition (2.10) must be met for some unique allotment $R_{c(z)}^{\alpha}(X) \in B(c)$. The measure of separation and compactness of the allotment $V_{MSC}(R_c^*(X);c)$ increases when c is closer to n. Thus, the optimum value of c is obtained by minimizing $V_{MSC}(R_c^*(X);c)$ over $c = c_{\min},\ldots,c_{\max}$ where $2 \le c_{\min}$ and $c_{\max} < n$. Note at this point that the measure of separation and compactness of the allotment (2.32) is an extended version of the measure of compactness of the allotment which was proposed in [141].

2.2.6 Experimental Results

Let us consider results of some numerical experiments for some well-known data sets. In the first place, let us consider results of relational clustering algorithms application to the Yang and Wu's two-dimensional data set which is presented in Figure 1.6.

Yang and Wu's data can be presented as a matrix of attributes $\hat{X}_{16\times 2} = [\hat{x}_i^{t_1}]$, $i = 1,\ldots,16$, $t_1 = 1,2$, in which $\hat{x}_i^{t_1}$ is the value of the t_1-th attribute for the i-th object. The data were preprocessed according to the formulae (2.11), (2.15) and (1.26). The D-AFC(c)-algorithm was applied to the obtained matrix of fuzzy tolerance for $c = 2,\ldots,5$.

By executing the D-AFC(c)-algorithm, we obtain that the number $c = 5$ of fuzzy clusters in the allotment sought $R_c^*(X)$ is suboptimal. The corresponding values of validity measures were calculated for $c = 2,\ldots,4$ and these validity measures are plotted in Figures 2.2 – 2.4.

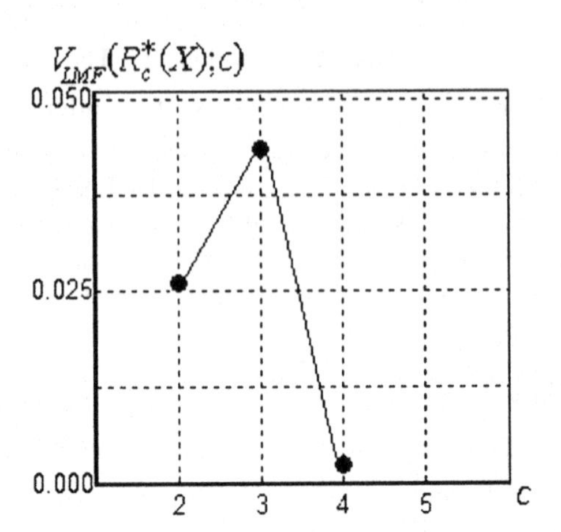

Fig. 2.2 Plot of the linear measure of fuzziness for Yang and Wu's data set

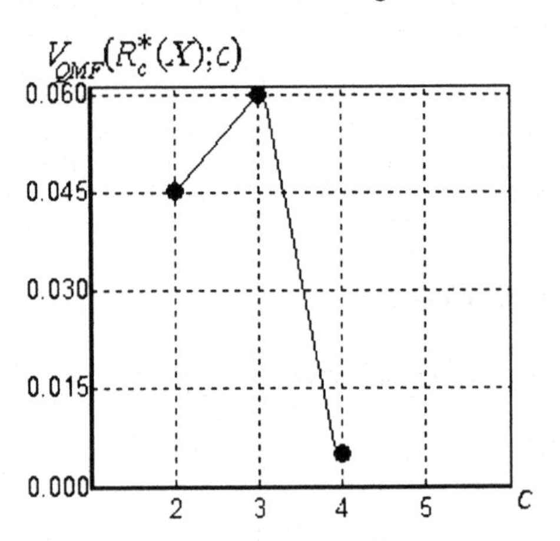

Fig. 2.3 Plot of the quadratic measure of fuzziness for Yang and Wu's data set

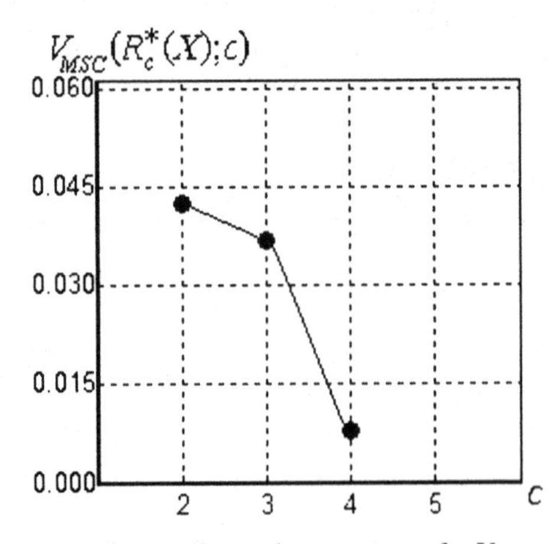

Fig. 2.4 Plot of the measure of separation and compactness for Yang and Wu's data set

The actual number of fuzzy clusters is equal 3 and this number corresponds to the maximum of the linear measure of fuzziness of the allotment (2.30) and the maximum of the quadratic measure of fuzziness of the allotment (2.31). The allotment $R^*_c(X)$ among three fully separated fuzzy clusters was obtained for the tolerance threshold $\alpha = 0.955$. The membership functions of the three classes of the allotment are presented in Figure 2.5, where the membership values of the first class are represented by ○, the membership values of the second class are represented by ■ and the membership values of the third class are represented by ▲.

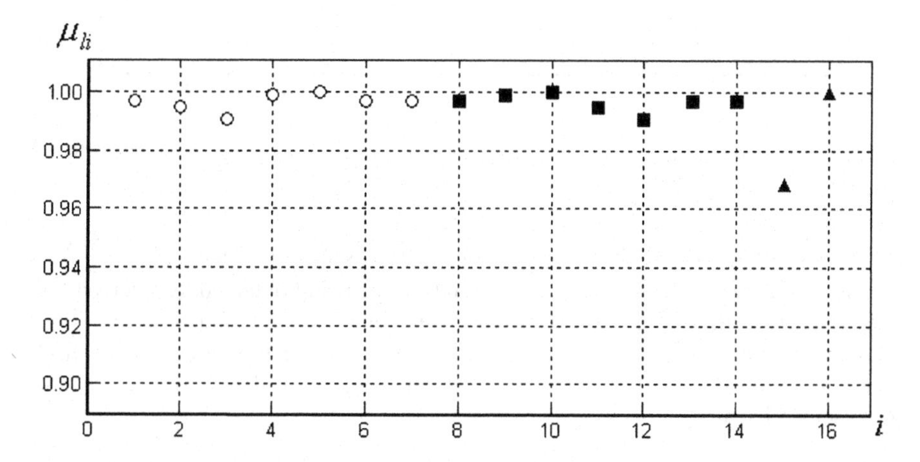

Fig. 2.5 Membership functions of three fuzzy clusters obtained from the D-AFC(c)-algorithm

The first class is formed by 7 elements, the second class is formed by 7 elements and the third class includes 2 elements. The fifth object is the typical point of the first fuzzy cluster, the tenth object is the typical point of the fuzzy cluster which corresponds to the second class, and the sixteenth object is the typical point of the fuzzy cluster which corresponds to the third class.

Let us consider a result of application of the D-AFC-PS(c)-algorithm to Yang and Wu's data set. The computational condition was determined as follows: the number of fuzzy clusters $c = 2$, the value of accuracy threshold $\varepsilon = 0.00001$, and the objects x_1 and x_9 were selected as labeled objects with their membership functions $y_{1(1)} = y_{2(9)} = 1.0$. The allotment $R^*_c(X)$ among two particularly separated fuzzy clusters was obtained for the tolerance threshold $\alpha = 0.92663$. Values of membership functions obtained from the D-AFC-PS(c)-algorithm are shown in Figure 2.6, where membership values of the first class are represented by ○ and membership values of the second class are represented by ■.

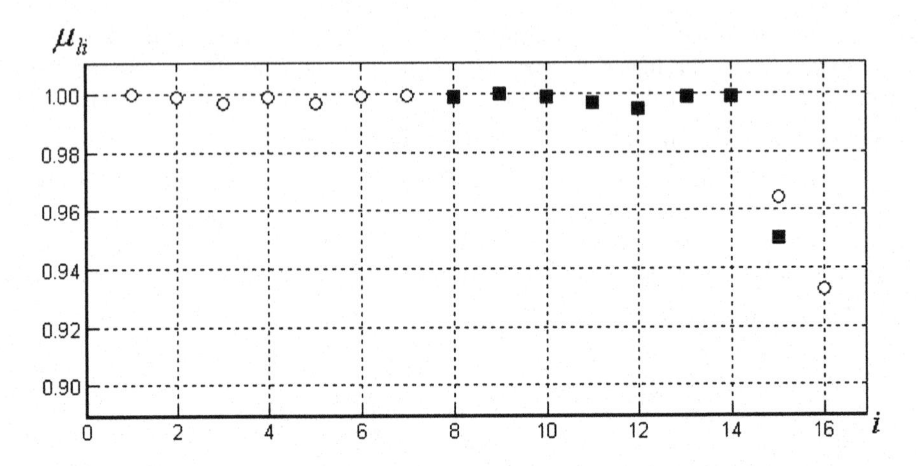

Fig. 2.6 Membership functions of two fuzzy clusters obtained by using the D-AFC-PS(c)-algorithm

By executing the D-PAFC-algorithm, the principal allotment $R_P^{0.94875}(X)$ among two fuzzy clusters, which corresponds to the result obtained, is received for the tolerance threshold $\alpha = 0.94875$. The membership functions of two classes are presented in Figure 2.7 in which the membership values of the first class are represented by ○, the membership values of the second class are represented by ■.

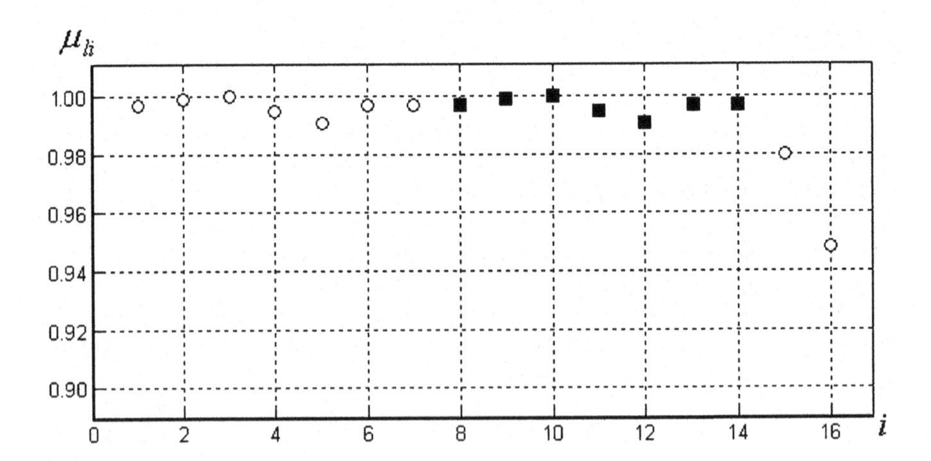

Fig. 2.7 Membership functions of two fuzzy clusters obtained by using the D-PAFC-algorithm

Thus, the third object is the typical point of the first fuzzy cluster and the tenth object is the typical point of the fuzzy cluster which corresponds to the second class.

Let us consider results btained by different clustering algorithms applied to Sneath and Sokal's [104] artificial data set shown in Figure 2.8.

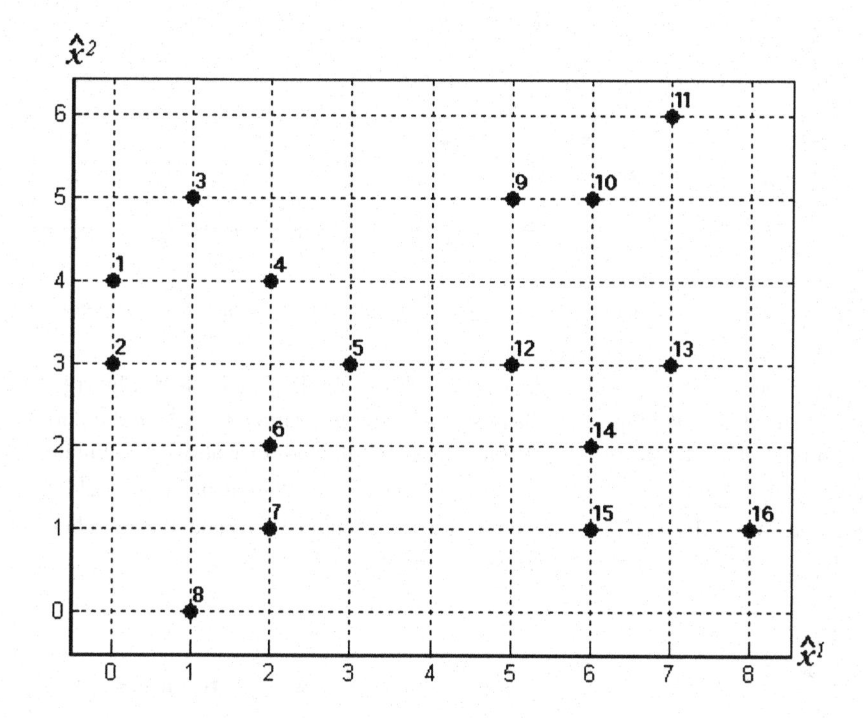

Fig. 2.8 Sneath and Sokal's data set

Bezdek [10] applied the FCM-algorithm to this data set for the weighting exponent value $\gamma = 2$. By computing some validity criteria for a variety of choices of c, $c = 2,\ldots,6$, Bezdek [10] infers that $c = 2$ is the most appropriate choice. The membership functions of two fuzzy clusters are shown in Figure 2.9. The membership values of the first class are represented by ○ and the membership values of the second class are represented by ■. The vectors $\tau^1 = (1.43, 2.82)$ and $\tau^2 = (6.14, 3.15)$ are prototypes of the corresponding fuzzy clusters $A^l \in P(X)$, $l = 1,2$, obtained by the FCM-algorithm.

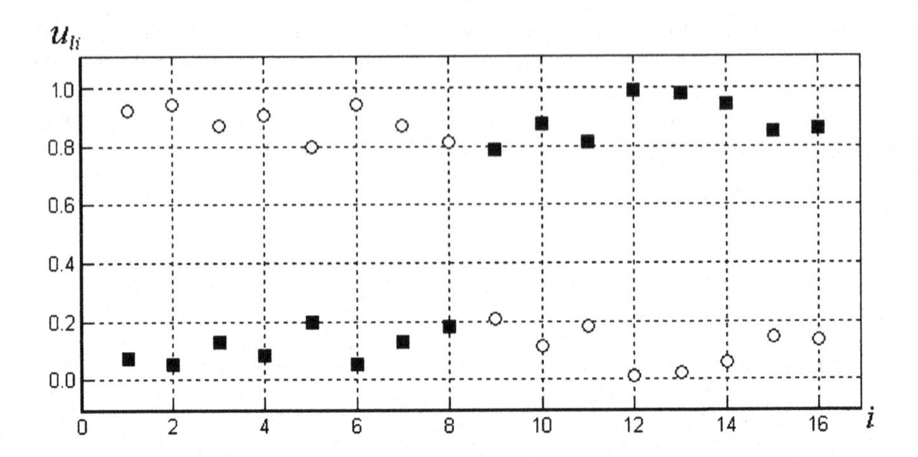

Fig. 2.9 Membership functions of two fuzzy clusters obtained by using the FCM-algorithm

For comparison, Li and Mukaidono [71] also applied their GCM-algorithm to this data set for the number of classes $c = 2$ and the value $\lambda = 1.5$ in the entropy term in (1.79). The fuzzy c -partition was also obtained. Probabilistic membership degrees u_{li}, $l = 1,2$, $i = 1,\ldots,16$, of two clusters are shown in Figure 2.10. The membership values of the first class are represented by ○ and the membership values of the second class are represented by ■.

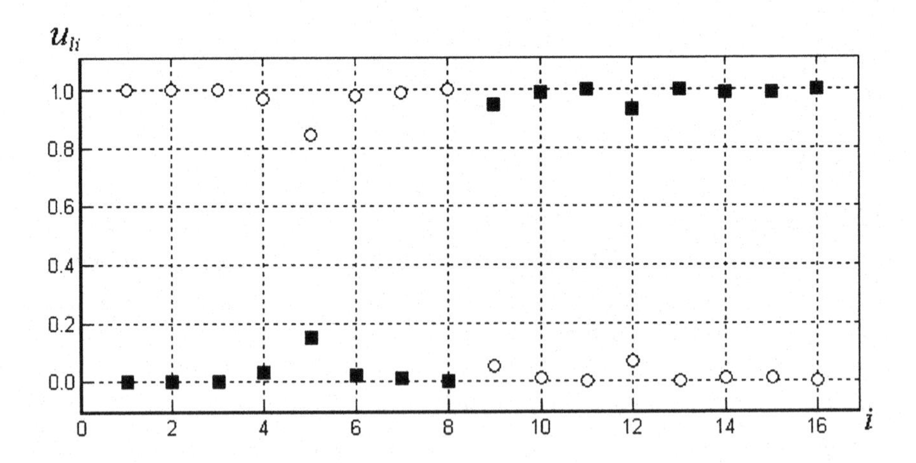

Fig. 2.10 Membership functions of two fuzzy clusters obtained by using the GCM-algorithm

The fuzzy clusters prototypes $\tau^1 = (1.41, 2.76)$ and $\tau^2 = (6.17, 3.24)$ were also obtained by the GCM-algorithm in [71].

In order to compare the algorithms with the proposed relational clustering algorithms, the matrix of fuzzy tolerance relation $T = [\mu_T(x_i, x_j)]$, $i, j = 1,\ldots,16$ was constructed according to formulae (2.11), (2.15) and (1.26). The D-AFC(c)-algorithm was applied to the matrix of fuzzy tolerance for $c = 2,\ldots,5$ using the validity measures (2.30), (2.31) and (2.32). The performance of the validity measures are shown in Figures 2.11 – 2.13.

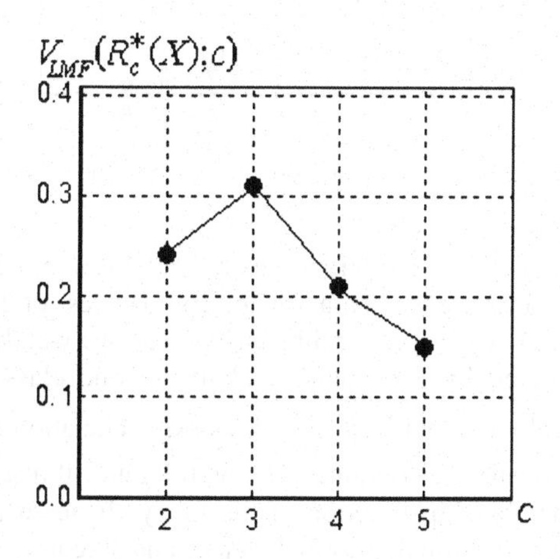

Fig. 2.11 Plot of the linear measure of fuzziness for Sneath and Sokal's data set

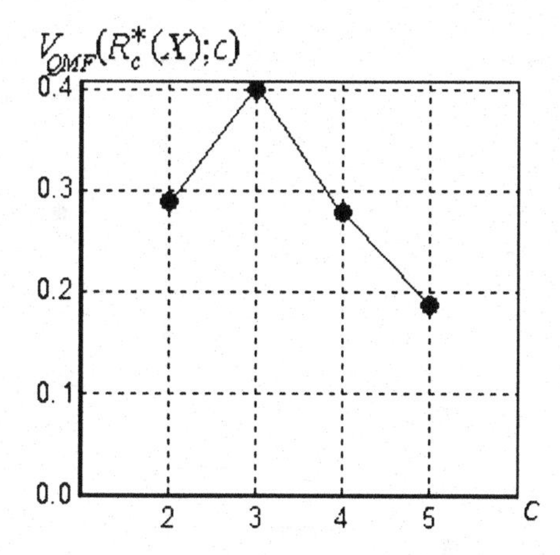

Fig. 2.12 Plot of the quadratic measure of fuzziness for Sneath and Sokal's data set

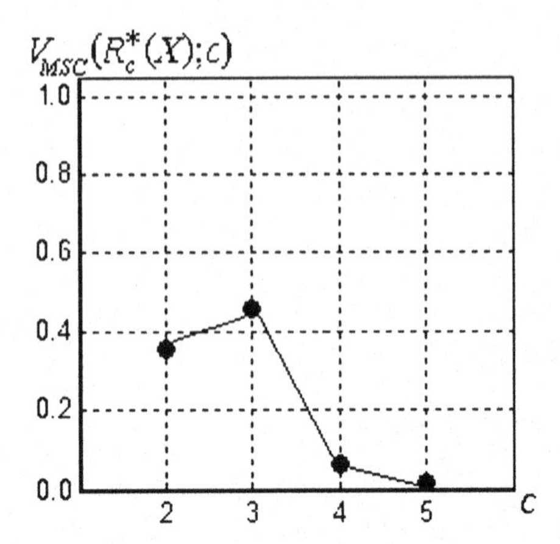

Fig. 2.13 Plot of the measure of separation and compactness for Sneath and Sokal's data set

The actual number of fuzzy clusters is equal 2 and this number corresponds to the first minimum of the measure of separation and compactness of the allotment (2.32).

By executing the D-AFC(c)-algorithm for two classes we obtain the following: the first class is formed by 8 elements and the second class is composed of 9 elements. The fifth element belongs to both classes. The allotment $R_c^*(X)$, which corresponds to that result, was obtained for the tolerance threshold $\alpha = 0.81944$. The value of the membership function of the fuzzy cluster which corresponds to the first class is maximal for the second object and is equal one. Therefore, the second object is the typical point of the first fuzzy cluster. The membership value

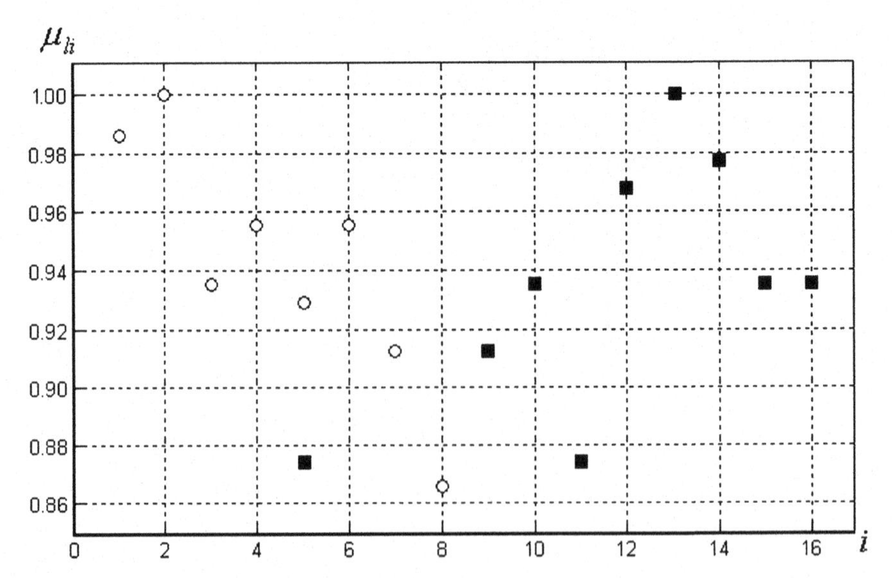

Fig. 2.14 Membership functions of two fuzzy clusters obtained by using the D-AFC(c)-algorithm

of the thirteenth object is equal one for the second fuzzy cluster. Thus, the thirteenth object is the typical point of the second fuzzy cluster. The membership functions of two classes of the allotment are presented in Figure 2.14 and values which equal zero are not shown in the figure.

Notice, that the Gaussian membership function is sharper than the membership function, which is obtained from the D-AFC(c)-algorithm, but the essential interpretation of the results which obtained are from the D-AFC(c)-algorithm is better than in the case of the GCM-algorithm.

Let us consider the result of experiment with the D-AFC-PS(c)-algorithm. The experiment was made for the set of labeled objects $X_L = \{x_8 = x_{L(1)}, x_{13} = x_{L(2)}\}$ with their membership functions $y_{1(8)} = 0.85351$ and $y_{2(13)} = 0.85916$. The membership derees $y_{l(j)}$, $l \in \{1,2\}$, $j \in \{8,13\}$, correspond to both labeled object $x_{L(j)} \in X_L$, $j \in \{8,13\}$, and were calculated according to the technique proposed in [127]. By executing the D-AFC-PS(c)-algorithm for two classes we obtain the result which is equal to the result obtained by the D-AFC(c)-algorithm.

Let us consider a result of application of the D-PAFC-algorithm to Sneath and Sokal's data set. By executing the D-PAFC-algorithm, the principal allotment $R_P^{0.91319}(X)$ among four fuzzy clusters, which corresponds to the result, is obtained for the tolerance threshold $\alpha = 0.91319$. The membership functions of four classes are presented in Figure 2.15 in which the membership values of the first class are represented by ○, the membership values of the second class are represented by ■, the membership values of the third class are represented by ▲, and the membership values of the fourth class are represented by □. The values which equal zero are not shown in the figure.

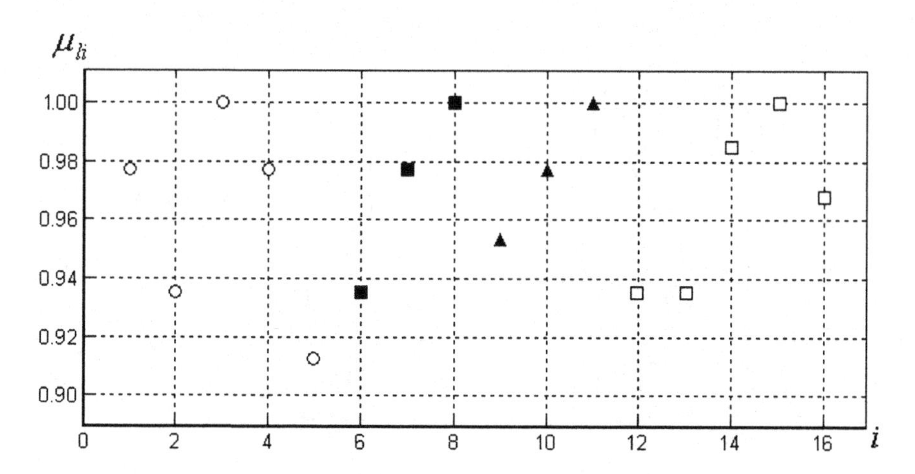

Fig. 2.15 Membership functions of four fuzzy clusters obtained by using the D-PAFC-algorithm

The value of the membership function of the fuzzy cluster which corresponds to the first class is maximal for the third object and is equal one. So, the third object is the typical point of the first fuzzy cluster. The membership value of the eighth object is equal one for the second fuzzy cluster. Thus, the eighth object is the typical point of the second fuzzy cluster. The membership value of the eleventh object is equal one for the third fuzzy cluster. That is why the eleventh object is the typical point of the third fuzzy cluster. The value of the membership function of the fuzzy cluster which corresponds to the fourth class is maximal for the fifteenth object and the object is the typical point of the corresponding cluster.

Let us consider an application of the relational clustering algorithms to the classification problem for the following illustrative example. The problem of classification of family portraits coming from three families was considered by Tamura, Higuchi and Tanaka [106]. The number of portraits was equal to 16 and the real portrait assignment among three classes is presented in Figure 2.16.

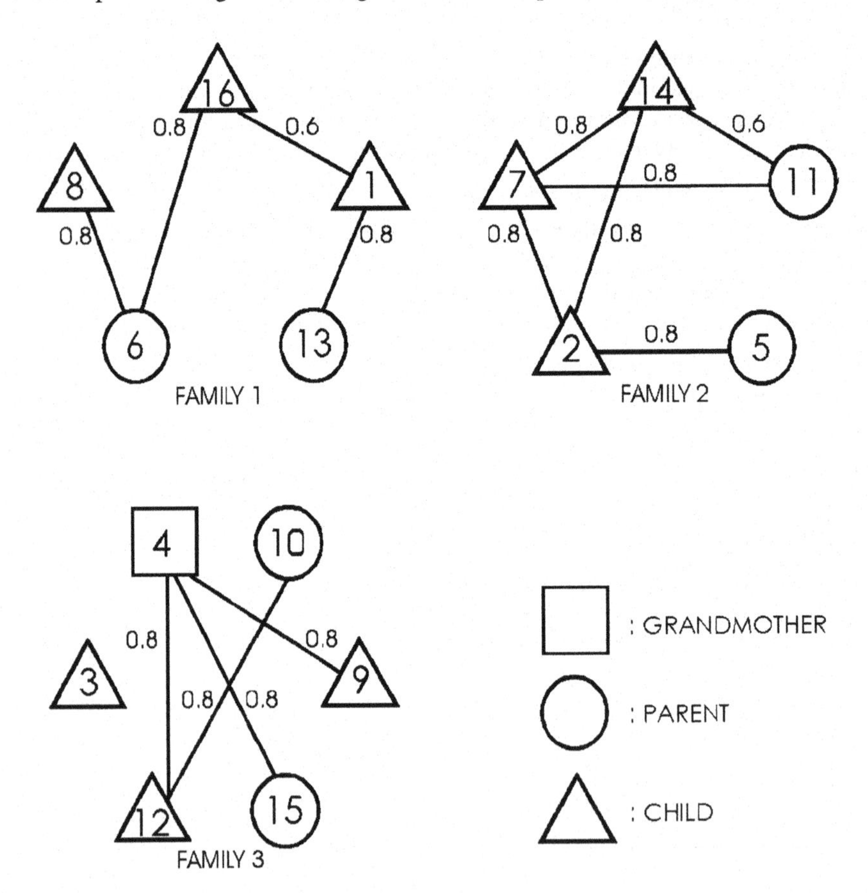

Fig. 2.16 Real portraits classification

The data were originally analyzed in order to identify families by using the technique of first transforming the matrix of a fuzzy tolerance into a matrix of a fuzzy similarity relation and then taking an appropriate α-level of the fuzzy similarity relation [106]. The best partition proved to be obtained with the α-level equal to 0.6. The partition identified the following three families $A^1 = \{x_1, x_6, x_8, x_{13}, x_{16}\}$, $A^2 = \{x_2, x_5, x_7, x_{11}, x_{14}\}$ and $A^3 = \{x_4, x_9, x_{10}, x_{12}, x_{15}\}$. However, person x_3 is not a member of any of the three families.

The subjective similarities assigned to the individual pairs of portraits collected in the tabular form are presented in Table 2.4.

Table 2.4 The matrix of subjective similarities

T	x_1	x_2	x_3	x_4	x_5	x_6	x_7	x_8	x_9	x_{10}	x_{11}	x_{12}	x_{13}	x_{14}	x_{15}	x_{16}
x_1	1.0															
x_2	0.0	1.0														
x_3	0.0	0.0	1.0													
x_4	0.0	0.0	0.4	1.0												
x_5	0.0	0.8	0.0	0.0	1.0											
x_6	0.5	0.0	0.2	0.2	0.0	1.0										
x_7	0.0	0.8	0.0	0.0	0.4	0.0	1.0									
x_8	0.4	0.2	0.2	0.5	0.0	0.8	0.0	1.0								
x_9	0.0	0.4	0.0	0.8	0.4	0.2	0.4	0.0	1.0							
x_{10}	0.0	0.0	0.2	0.2	0.0	0.0	0.2	0.0	0.2	1.0						
x_{11}	0.0	0.5	0.2	0.2	0.0	0.0	0.8	0.0	0.4	0.2	1.0					
x_{12}	0.0	0.0	0.2	0.8	0.0	0.0	0.0	0.0	0.4	0.8	0.0	1.0				
x_{13}	0.8	0.0	0.2	0.4	0.0	0.4	0.0	0.4	0.0	0.0	0.0	0.0	1.0			
x_{14}	0.0	0.8	0.0	0.2	0.4	0.0	0.8	0.0	0.2	0.2	0.6	0.0	0.0	1.0		
x_{15}	0.0	0.0	0.4	0.8	0.0	0.2	0.0	0.0	0.2	0.0	0.0	0.2	0.2	0.0	1.0	
x_{16}	0.6	0.0	0.0	0.2	0.2	0.8	0.0	0.4	0.0	0.0	0.0	0.0	0.4	0.2	0.0	1.0

The D-AFC(c)-algorithm was applied to the matrix of fuzzy tolerances for $c = 2,\ldots,5$ using the validity measures (2.30) – (2.32). The performance of the proposed validity measures is shown in Figures 2.17 – 2.19.

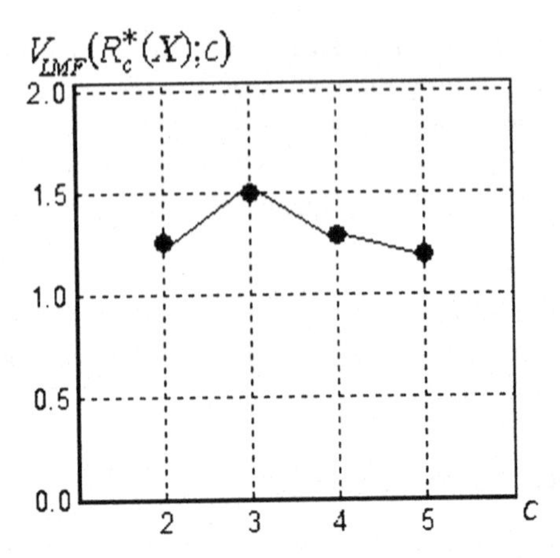

Fig. 2.17 Plot of the linear measure of fuzziness for Tamura's data set

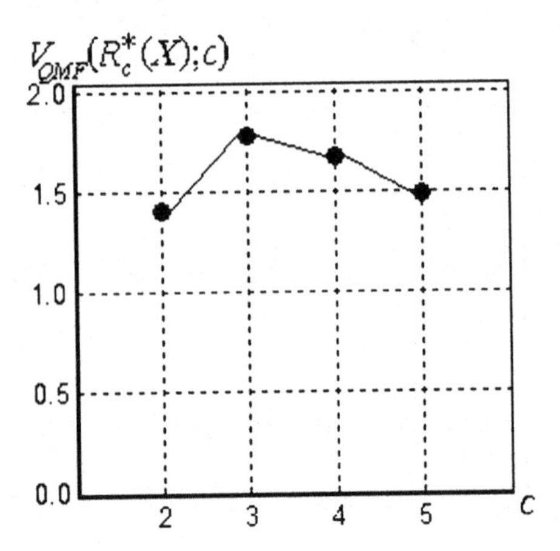

Fig. 2.18 Plot of the quadratic measure of fuzziness for Tamura's data set

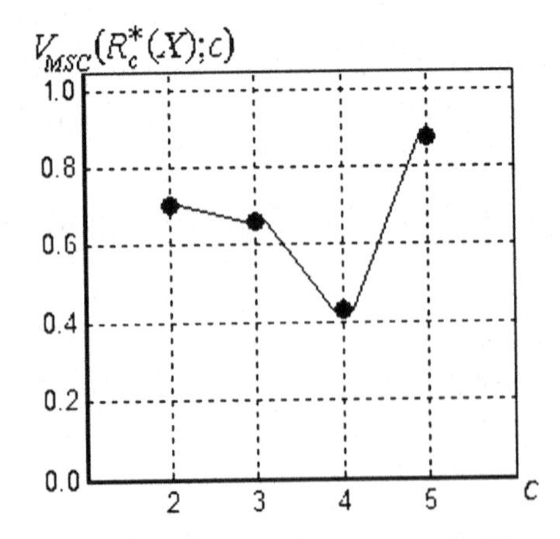

Fig. 2.19 Plot of the measure of separation and compactness for Tamura's data set

The actual number of fuzzy clusters is equal 3 and the number corresponds to the maximum of the linear measure of fuzziness of the allotment (2.30) and the maximum of the quadratic measure of fuzziness of the allotment (2.31). On the other hand, the minimal value of the measure of separation and compactness of the allotment is equal 0.4385 and the value corresponds to the four fully separate fuzzy clusters. The value of the total number of elements in the intersection areas is equal 3 for the allotment among two particularly separated fuzzy clusters, for $c = 3$ we have the total number of elements in intersection areas equal 2 and the value of the total number of elements in intersection areas is equal 1 for $c = 5$. The membership functions of three classes of the allotment $R_c^*(X)$ are presented in Figure 2.20 in which the membership values of the first class are represented by ○, the membership values of the second class are represented by ■ and the membership values of the third class are represented by ∇ .

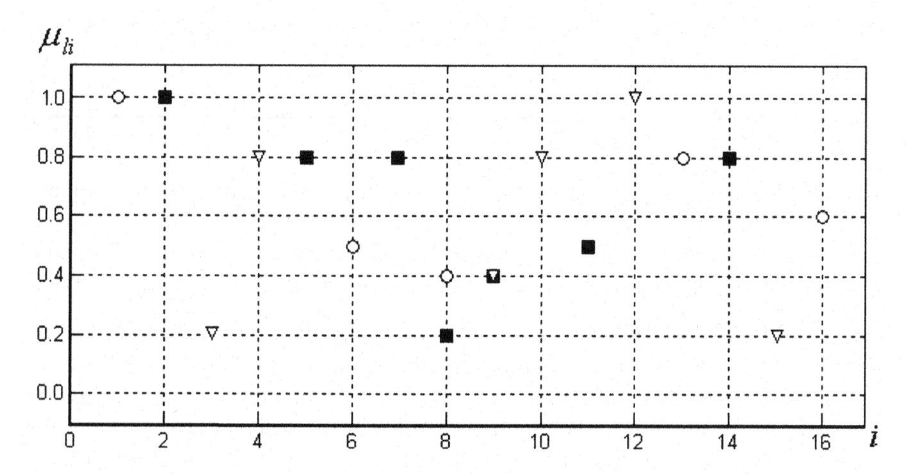

Fig. 2.20 Membership functions of three fuzzy clusters obtained by the D-AFC(c)-algorithm

By executing the D-AFC(c)-algorithm for three classes, we obtain the allotment $R_c^*(X)$ among particularly separated fuzzy clusters which corresponds to the result that is obtained for the tolerance threshold $\alpha = 0.2$. The ninth element of the set of objects is belonging to the second class and to the third class and membership values are equal, $\mu_{2\,9} = \mu_{3\,9} = 0.4$.

Let us consider the result obtained by the D-AFC-PS(c)-algorithm. The experiment was made for the set of labeled objects $X_L = \{x_{13} = x_{L(1)}, x_5 = x_{L(2)}, x_{10} = x_{L(3)},\}$ with their membership grades $y_{1(13)} = 1.0$, $y_{2(5)} = 1.0$, and $y_{3(10)} = 1.0$. Thus, we obtain the allotment $R_c^*(X)$ among three particularly separated fuzzy clusters, which corresponds to the result, which are obtained for the tolerance threshold $\alpha = 0.2$. The values of membership functions obtained by the D-AFC-PS(c)-algorithm are shown in Figure 2.21 in which the membership values of the first class are represented by ○, the membership values of the second class are represented by ■, and the membership values of the third class are represented by ∇.

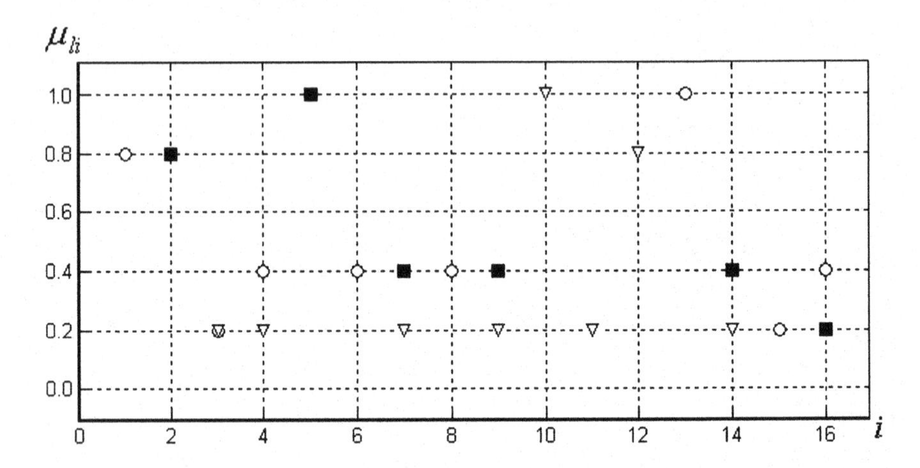

Fig. 2.21 Membership functions of fuzzy clusters obtained by the D-AFC-PS(c)-algorithm

Let us consider the result of application of the D-PAFC-algorithm to Tamura's data set. After application of the D-PAFC-algorithm to the matrix of the initial data, the principal allotment $R_P^{0.4}(X)$ among four fuzzy clusters, which corresponds to the result, is obtained for the tolerance threshold $\alpha = 0.4$. By executing the D-PAFC-algorithm we obtain the following: the first class is composed of 5 elements, all belonging to Family 1; the second class is formed by 6 elements, with five elements corresponding to Family 2 and one element corresponding to Family 3; the third class consists of 3 elements, all belonging to Family 3, and the fourth class contains 2 elements, all from Family 3. So, the union of the third and fourth classes is the class which corresponds to Family 3 and there is one classification error . The ninth element of the set of objects is the misclassified object.The membership functions of four classes are presented in Figure 2.22 in which the membership values of the first class are represented by ○, the membership values of the second class are represented by ■, the membership values of the third class are represented by ∇, and the membership values of the fourth class are represented by □.

The sixth object is the typical point of the first fuzzy cluster, the second object is the typical point of the second fuzzy cluster, the fifteenth object is the typical point of the third fuzzy cluster, and the tenth object is the typical point of the fourth fuzzy cluster.

In order to compare the presented relational clustering algorithms with the well-known relational fuzzy clustering NERFCM-algorithm, we transformed the initial matrix into the dissimilariies matrix by taking the complement of the relationship

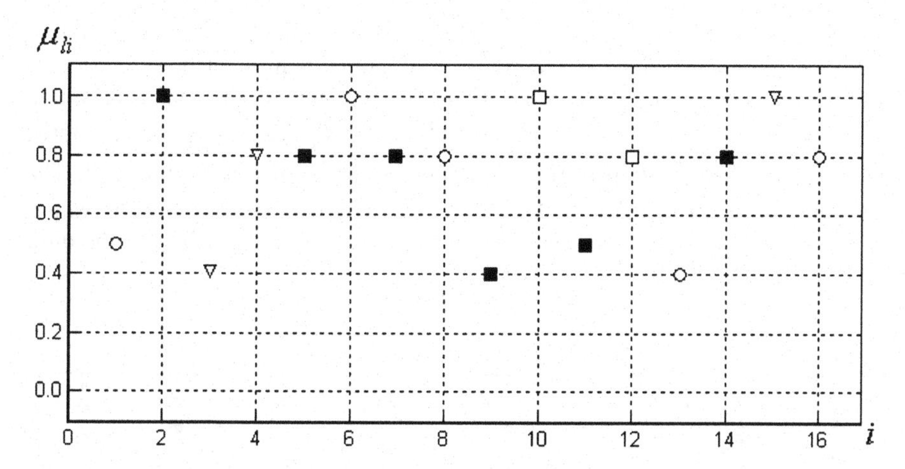

Fig. 2.22 Membership functions of four fuzzy clusters obtained by the D-PAFC-algorithm

degrees. The membership values originating from the NERFCM-algorithm are presented in [135]. The fuzzy c-partition $P(X)$ is produced by the NERFCM-algorithm. So, condition (1.71) is met for the membership values u_{li}, $0 \le u_{li} \le 1$, $1 \le l \le c$, $1 \le i \le n$, originating from the NERFCM-algorithm. If the maximum memberships rule (1.96) is applied to the matrix $P_{3 \times 16} = [u_{li}]$, $l = 1, \ldots, c$, $i = 1, \ldots, n$, of the fuzzy c-partition $P(X)$, then the result of the application of the NERFCM-algorithm to Tamura's data is similar to the real portraits classification. On the other hand, the ARCA-algorithm [25] was applied to Tamura's data using the compactness and separation index (1.91), for $c = 2, \ldots, 5$. The performance of the compactness and separation index is shown in Figure 2.23.

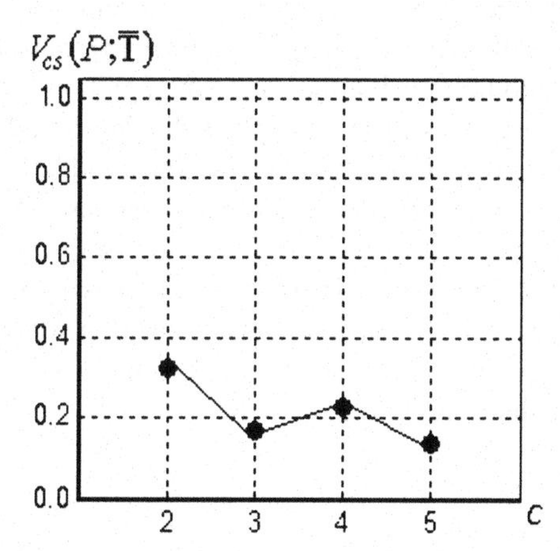

Fig. 2.23 Plot of the compactness and separation index as a function of the number of clusters

Thus, we observed that the minimal value of the compactness and separation index corresponds to five fuzzy clusters. The optimal number of fuzzy clusters is equal 3 and this number corresponds to the first minimum of the compactness and separation index (1.91). The membership values u_{li}, $l = 1,2$, $i = 1,\ldots,16$, of three clusters of the fuzzy c-partition $P(X)$ obtainedare shown in Figure 2.24. The membership values of the first class are represented by ○, the membership values of the second class are represented by ■, and the membership values of the third class are represented by ∇.

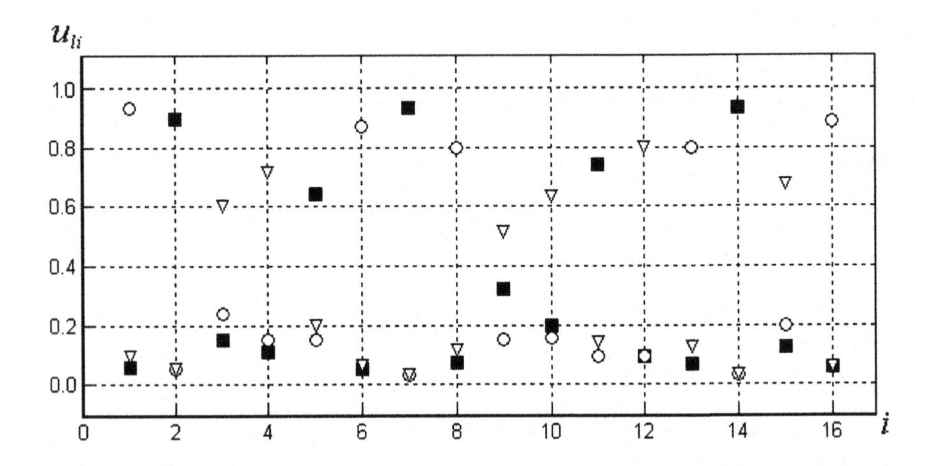

Fig. 2.24 Membership functions of three fuzzy clusters obtained by the ARCA-algorithm

In general, the result obtained by usingthe ARCA-algorithm is similar to the result obtained by the NERFCM-algorithm.

Let us consider the application of the relational clustering algorithms to the classification problem for Anderson's Iris data set [3]. The Iris database is the most known database to be found in the pattern recognition literature. The data set represents different categories of the Iris plants having four attribute values. The four attribute values represent the sepal length, sepal width, petal length and petal width measured for 150 iris species. It has three classes: Setosa, Versicolor and Virginica, with 50 samples per class. The problem is to classify the plants into three subspecies on the basis of this information. It is known that two classes, Versicolor and Virginica, have some amount of overlap while the Setosa class is linearly separable from the other two ones. The real assignments to the three classes are presented in Table 2.5.

Table 2.5 Real object assignments in the Iris data set

Class		Numbers of objects
Number	Name	
1	SETOSA	1, 6, 10, 18, 26, 31, 36, 37, 40, 42, 44, 47, 50, 51, 53, 54, 55, 58, 59, 60, 63, 64, 67, 68, 71, 72, 78, 79, 87, 88, 91, 95, 96, 100, 101, 106, 107, 112, 115, 124, 125, 134, 135, 136, 138, 139, 143, 144, 145, 149
2	VERSICOLOR	3, 8, 9, 11, 12, 14, 19, 22, 28, 29, 30, 33, 38, 43, 48, 61, 65, 66, 69, 70, 76, 84, 85, 86, 92, 93, 94, 97, 98, 99, 103, 105, 109, 113, 114, 116, 117, 118, 119, 120, 121, 128, 129, 130, 133, 140, 141, 142, 147, 150
3	VIRGINICA	2, 4, 5, 7, 13, 15, 16, 17, 20, 21, 23, 24, 25, 27, 32, 34, 35, 39, 41, 45, 46, 49, 52, 56, 57, 62, 73, 74, 75, 77, 80, 81, 82, 83, 89, 90, 102, 104, 108, 110, 111, 122, 123, 126, 127, 131, 132, 137, 146, 148

The matrix of fuzzy tolerance relation $T = [\mu_T(x_i, x_j)]$, $i, j = 1,\ldots,150$, was constructed according to the formulae (2.11), (2.15) and (1.26). The D-AFC(c)-algorithm was applied to the matrix of fuzzy tolerance relation for $c = 2,\ldots,5$ using the validity measures (2.30), (2.31) and (2.32). The value of the accuracy threshold was selected as $\varepsilon = 0.00001$. The performance of the validity measures are shown in Figures 2.25 – 2.27.

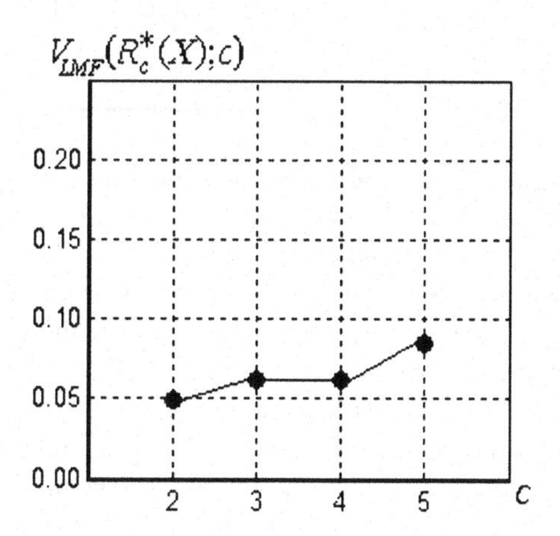

Fig. 2.25 Plot of the linear measure of fuzziness for Anderson's Iris data set

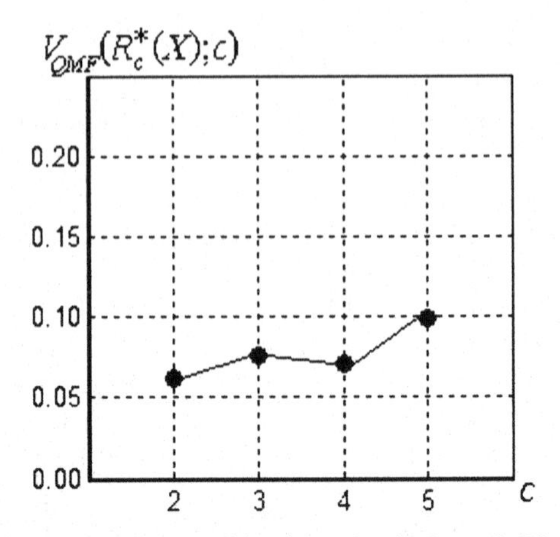

Fig. 2.26 Plot of the quadratic measure of fuzziness for Anderson's Iris data set

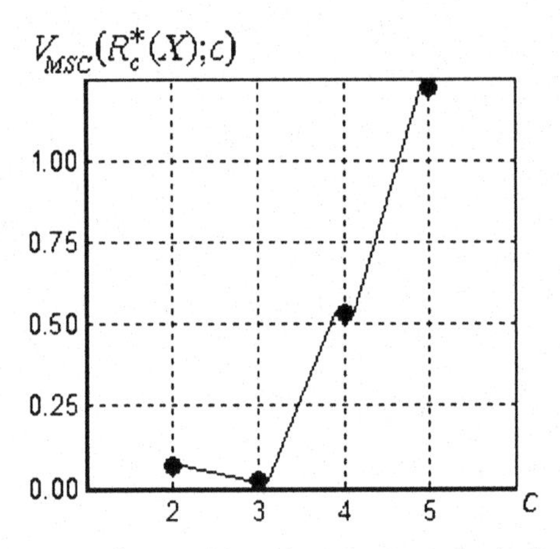

Fig. 2.27 Plot of the measure of separation and compactness for Anderson's Iris data set

By executing the D-AFC(c)-algorithm for $c = 2,\ldots,5$, we obtain that the optimal cluster number is chosen at $c = 5$ for the linear measure of fuzziness of the allotment and the quadratic measure of fuzziness of the allotment. However, the number of fuzzy clusters $c = 3$ corresponds to the first maximum for both validity measures. The measure of separation and compactness of the allotment finds the optimal cluster number at $c = 3$. The value of the total number of elements in intersection areas is equal to 10 for the allotment among four particularly separate fuzzy clusters and the value of the total number of elements in intersection areas is equal to 18 for $c = 5$. So, the results of the proposed validity measures seem to be appropriate. For the comparison purpose, the Iris data set was also tested by the FCM-algorithm using the compactness and separation index (1.91). The FCM-algorithm was applied to the data set with the weighting exponent $\gamma = 2.0$ and the value of a small threshold $\varepsilon = 0.001$. The performance of the compactness and separation index for the data set is shown in Figure 2.28.

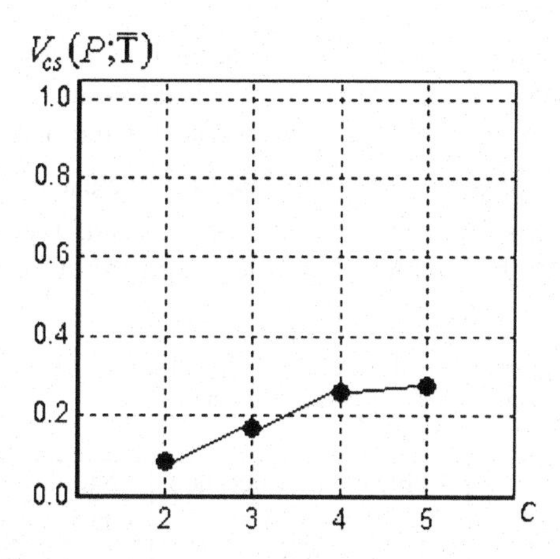

Fig. 2.28 Plot of the compactness and separation index as a function of the number of clusters

The optimal cluster number is chosen at $c = 2$ for the compactness and separation index. Note that most validity measures reported in the literature provide two clusters for these data [13].On the other hand, Gath and Geva [39] applied their GG-algorithmto the Iris data set for $c = 2,\ldots,6$. The fuzzy hypervolume (1.92) and the partition density (1.94) were used in the experiments and the optimal cluster number c is chosen at $c = 3$ for both validity measures

Three experiments were performed for the number of fuzzy clusters $c = 3$, the value of accuracy threshold $\varepsilon = 0.00001$ and different distances (2.13) – (2.15). The results obtained are summarized in Table 2.6.

Table 2.6 Results of application of the D-AFC(c)-algorithm to Anderson's Iris data for different distances

Distance	Main characteristics of the result of classification		
	Number of misclassified objects	Value of the tolerance threshold	Typical points of fuzzy clusters in the obtained allotment
The normalized Hamming distance (2.13)	15	$\alpha = 0.81916$	$x_{95} = \tau^1$, $x_{98} = \tau^2$, $x_{126} = \tau^3$
The normalized Euclidean distance (2.14)	6	$\alpha = 0.81044$	$x_{95} = \tau^1$, $x_{98} = \tau^2$, $x_{23} = \tau^3$
The squared normalized Euclidean distance (2.15)	6	$\alpha = 0.96418$	$x_{95} = \tau^1$, $x_{98} = \tau^2$, $x_{73} = \tau^3$

Four experiments were made for the number of fuzzy clusters $c = 3$ and different values of the accuracy threshold ε . The formula (2.11) and the squared normalized Euclidean distance (2.15) were used for the data preprocessing. The allotment $R_c^*(X)$ among three fully separated fuzzy clusters was obtained in each experiment. The results for Anderson's Iris data set using the D-AFC(c)-algorithm for different values of the accuracy threshold ε are presented in the Table 2.7.

Table 2.7 Results classification of Anderson's Iris data set obtained by the D-AFC(c)-algorithm for different values of the accuracy threshold

Value of the accuracy threshold, ε	Value of the tolerance threshold, α	Number of typical points		
		Class 1	Class 2	Class 3
ε=0.01	α=0.97	42	15	12
ε=0.001	α=0.965	7	2	1
ε=0.0001	α=0.9642	2	1	1
ε=0.00001	α=0.96418	1	1	1

The object assignments resulting from the application of the D-AFC(c)-algorithm to Anderson's Iris data in each experiment are presented in Table 2.8.

Table 2.8 Results of the D-AFC(c)-algorithm application: the object assignments

Class		Numbers of objects
Number	Name	
1	SETOSA	1, 6, 10, 18, 26, 31, 36, 37, 40, 42, 44, 47, 50, 51, 53, 54, 55, 58, 59, 60, 63, 64, 67, 68, 71, 72, 78, 79, 87, 88, 91, 95, 96, 100, 101, 106, 107, 112, 115, 124, 125, 134, 135, 136, 138, 139, 143, 144, 145, 149
2	VERSICOLOR	3, **5**, 8, 11, 12, 14, 19, 22, **25**, 28, 29, 30, 33, 38, 43, 48, **56**, 61, 65, 66, 69, 70, 76, 84, 85, 86, **90**, 92, 93, 94, 97, 98, 99, 103, 105, 109, 113, 114, 116, 117, 118, 119, 120, 121, 128, 129, 130, 133, 140, 141, 142, 147, 150
3	VIRGINICA	2, 4, 7, **9**, 13, 15, 16, 17, 20, 21, 23, 24, 27, 32, 34, 35, 39, 41, 45, 46, 49, 52, 57, 62, 73, 74, 75, 77, 80, 81, 82, 83, 89, 102, 104, 108, 110, 111, 122, 123, 126, 127, 131, 132, 137, 146, **147**, 148

Misclassified objects are distinguished in Table 2.8. There are six classification errors. The membership functions μ_{li} of three fuzzy clusters of the allotment $R_c^*(X)$ which was obtained for the value of the accuracy threshold $\varepsilon = 0.00001$ are presented in Figure 2.29.

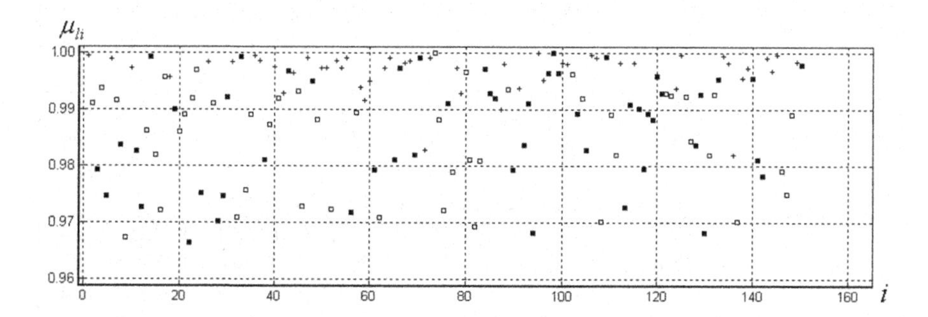

Fig. 2.29 Membership functions of three fuzzy clusters obtained by the D-AFC(c)-algorithm

The membership values which equal zero are not shown in this figure. The membership values of the first fuzzy cluster are represented by +, the membership values of the second fuzzy cluster are represented by ■, and the membership values of the third fuzzy cluster are represented by □.

The result obtained by the D-AFC-PS(c)-algorithm was described by Damaratski and Juodelis [28]. The formula (2.11) and the squared normalized Euclidean distance (2.15) were also used in the experiment. The conditions assumed were as follows: the number of fuzzy clusters $c = 3$, the value of accuracy threshold $\varepsilon = 0.0001$, and the set of labeled objects $X_L = \{x_{95} = x_{L(1)}, x_3 = x_{L(2)}, x_5 = x_{L(3)},\}$ with their membership grades $y_{1(95)} = y_{2(3)} = y_{3(5)} = 0.9929$. The allotment $R_c^*(X)$ among three partially separated fuzzy clusters was obtained by the D-AFC-PS(c)-algorithm for the value of tolerance threshold $\alpha = 0.9715$. The object assignments resulting from the use of the D-AFC-PS(c)-algorithm to Anderson's Iris data are presented in Table 2.9. The misclassified objects are distinguished in the table.

Table 2.9 Results of the application of the D-AFC-PS(c)-algorithm: the object assignments

Class		Numbers of objects
Number	Name	
1	SETOSA	1, 6, 10, 18, 26, 31, 36, 37, 40, 42, 44, 47, 50, 51, 53, 54, 55, 58, 59, 60, 63, 64, 67, 68, 71, 72, 78, 79, 87, 88, 91, 95, 96, 100, 101, 106, 107, 112, 115, 124, 125, 134, 135, 136, 138, 139, 143, 144, 145, 149
2	VERSICOLOR	3, 8, 11, 14, 19, **25**, 29, 30, 33, 38, 43, 48, 61, 65, 66, 69, 70, 76, 84, 85, 86, **90**, 92, 93, 97, 98, 99, 103, 105, 109, 114, 116, 117, 118, 119, 120, 121, 128, 129, 133, 140, 141, 142, 150
3	VIRGINICA	2, 4, 5, 7, **9**, **12**, 13, 15, 16, 17, 20, 21, **22**, 23, 24, 27, **28**, 32, 34, 35, 39, 41, 45, 46, 49, 52, 56, 57, 62, 73, 74, 75, 77, 80, 81, 82, 83, 89, **94**, 102, 104, 108, 110, 111, **113**, 122, 123, 126, 127, **130**, 131, 132, 137, 146, **147**, 148

There are ten classification error. The membership functions of three classes are presented in Figure 2.30.

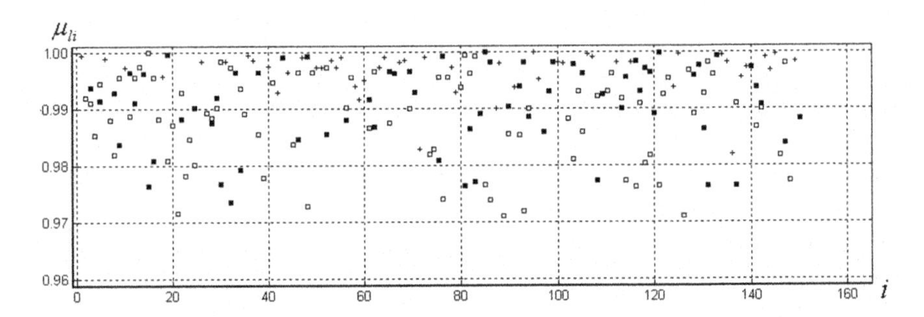

Fig. 2.30 Membership functions of three clusters obtained by the D-AFC-PS(c)-algorithm

Thus, the D-AFC-PS(c)-algorithm is sensitive to the selection of the set of labeled objects X_L and the estimation of membership derees $y_{l(j)}$, $l = 1,\ldots,c$, $j = 1,\ldots,c$, which are correspond to each labeled object $x_{L(j)} \in X_L$, $j = 1,\ldots,c$.

Let us consider the result of application of the D-PAFC-algorithm to Anderson's Iris data set. By executing the D-PAFC-algorithm, the principal allotment $R_P^{0.91655}(X)$ among two fuzzy clusters, which corresponds to the result, is obtained for the tolerance threshold $\alpha = 0.91655$. The seventynineth object is the typical point of the first fuzzy cluster and the sixtysecond object is the typical point of the fuzzy cluster which corresponds to the second class.

The object assignments obtained by the D-PAFC-algorithm in the experiment are presented in Table 2.10.

Table 2.10 Results of the D-PAFC-algorithm application: the object assignments

Class		Numbers of objects
Number	Name	
1	SETOSA	1, 6, 10, 18, 26, 31, 36, 37, 40, 42, 44, 47, 50, 51, 53, 54, 55, 58, 59, 60, 63, 64, 67, 68, 71, 72, 78, 79, 87, 88, 91, 95, 96, 100, 101, 106, 107, 112, 115, 124, 125, 134, 135, 136, 138, 139, 143, 144, 145, 149
2	VERSICOLOR&VIRGINICA	2, 3, 4, 5, 7, 8, 9, 11, 12, 13, 14, 15, 16, 17, 19, 20, 21, 22, 23, 24, 25, 27, 28, 29, 30, 32, 33, 34, 35, 38, 39, 41, 43, 45, 46, 48, 49, 52, 56, 57, 61, 62, 65, 66, 69, 70, 73, 74, 75, 76, 77, 80, 81, 82, 83, 84, 85, 86, 89, 90, 92, 93, 94, 97, 98, 99, 102, 103, 104, 105, 108, 109, 110, 111, 113, 114, 116, 117, 118, 119, 120, 121, 122, 123, 126, 127, 128, 129, 130, 131, 132, 133, 137, 140, 141, 142, 146, 147, 148, 150

The membership functions of two classes are presented in Figure 2.31. The membership values of the first fuzzy cluster are represented by + and the membership values of the second fuzzy cluster are represented by ■.

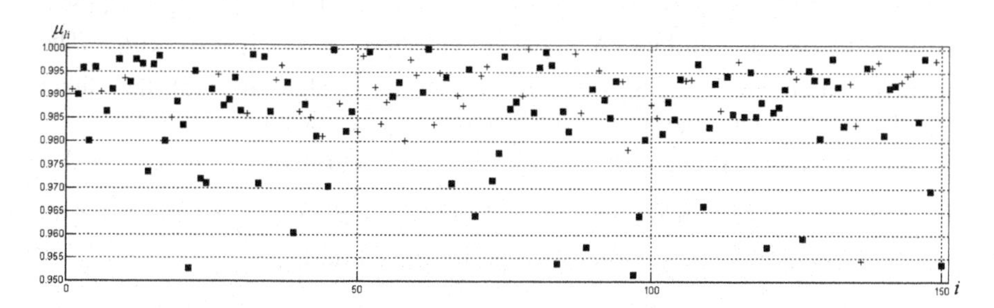

Fig. 2.31 Membership functions of two clusters obtained by the D-PAFC-algorithm

The FCM-algorithm of fuzzy clustering was applied to the data with the number of classes $c = 3$ and the weighting exponent $\gamma = 2.0$. The FCM-algorithm leads to the $15/150$ error rate. The PCM-algorithm of possibilistic clustering was applied to the data with the number of classes $c = 3$ and the noise distance $\eta_l = 0.1$ and $\eta_l = 0.3$, for all $l = 1,\ldots,c$. For the PCM-algorithm, the best result, equal to $8/150$, was obtained with $\eta_l = 0.1$, $l = 1,\ldots,c$. Łęski [74] applied the εPCM-algorithm to Anderson's Iris data with the insensitivity interval $\varepsilon = 0.1$, the noise distance $\eta_l = 0.5$, $l = 1,\ldots,c$, and the number of classes $c = 3$. So, the best result for the method, equal to $6/150$, was obtained with these values of parameters. The confusion matrices for three classes are presented in Table 2.11. The maximal membership rule (1.96) for the final hard partition matrix was used in each case.

Table 2.11 Results of numerical experiments: the confusion matrices

The FCM-algorithm with $\gamma = 2.0$			The PCM-algorithm with $\eta_l = 0.1$			The εPCM-algorithm with $\varepsilon = 0.1, \eta_l = 0.5$		
50	0	0	50	0	0	50	0	0
0	47	3	1	43	6	0	46	4
0	12	38	0	1	49	0	2	48

The results of numerical experiments seem to be satisfactory. Moreover, the membership functions obtained from the proposed relational clustering algorithms are sharper than the membership functions obtained from objective function based fuzzy clustering algorithms.

2.3 Prototype-Based Clustering Algorithms

Prototype-based heuristic possibilistic clustering algorithms are based on the transitive closure $\tilde{T}$ of fuzzy tolerance T and these algorithms can only be applied to the matrix of attributes $\hat{X}_{n \times m_1} = [\hat{x}_i^{t_1}]$, $i = 1,\ldots,n$, $t_1 = 1,\ldots,m_1$. The prototype-based clustering procedures include three algorithms of the direct classification and one algorithm of the hierarchical classification.

The matrix of attributes $\hat{X}_{n \times m_1} = [\hat{x}_i^{t_1}]$ is the matrix of initial data, and a distance $d(x_i, x_j)$ for fuzzy sets given by (2.13) – (2.15) is a parameter for all prototype-based heuristic algorithm of possibilistic clustering.

2.3.1 Direct Clustering Algorithms

The family of direct prototype-based clustering procedures which were proposed in [126] includes:

- D-AFC-TC-algorithm: using the construction of the allotment among an unknown number c of fully separated fuzzy clusters;
- D-PAFC-TC-algorithm: using the construction of the principal allotment among an unknown minimal number c of fully separated fuzzy clusters;
- D-AFC-TC(α)-algorithm: using the construction of the allotment among an unknown number c of fully separated fuzzy clusters with respect to the minimal value α of the tolerance threshold.

The unique allotment among an unknown number c of fuzzy clusters can be selected from the set of allotments which depends on the tolerance threshold. An idea of a leap in similarity values for finding an appropriate value α_ℓ of the tolerance threshold can be useful for the purpose. That is why an appropriate value α_ℓ of the tolerance threshold must be found in the ordered sequence $0 < \alpha_0 < \alpha_1 < \ldots < \alpha_\ell < \ldots < \alpha_Z \leq 1$. For this purpose, a leap heuristic was proposed in [122].

The leap heuristic can be described as a three-step procedure:

1. The values $g_\ell = \alpha_{\ell-1}/\alpha_\ell$ should be computed for all α_ℓ, $\ell = 1,\ldots,Z-1$, in the ordered sequence $0 < \alpha_0 < \alpha_1 < \ldots < \alpha_\ell < \ldots < \alpha_Z \leq 1$;
2. The value α_{ℓ^*} which corresponds to ℓ^* for some $g_{\ell^*} = \min\limits_{\ell} g_\ell$, $\ell = 1,\ldots,Z-1$, should be selected;
3. If multiple minimal g_{ℓ^*} values are obtained in the set $\{g_\ell\}$, $\ell = 1,\ldots,Z-1$, then the value α_{ℓ^*}, which corresponds to the minimal value of the index ℓ, should be selected.

The leap heuristic is used in the D-AFC-TC-algorithm and in the D-AFC-TC(α)-algorithm. The detection of the unique allotment $R_c^*(X)$ among an unknown number c of fully separated fuzzy clusters is the essence of the D-AFC-TC-algorithm.

The D-AFC-TC-algorithm is a nine-step classification procedure [126].

The D-AFC-TC-Algorithm

1. Construct the matrix of the fuzzy tolerance relation $T_{n \times n} = [\mu_T(x_i, x_j)]$ by normalizing the initial data $\hat{X}_{n \times m_1} = [\hat{x}_i^{t_1}]$, $i = 1,\ldots,n$, $t_1 = 1,\ldots,m_1$, and by choosing a suitable distance for fuzzy sets;
2. Construct the transitive closure $\tilde{T}$ of the fuzzy tolerance T according to the definition 1.24;

3. Construct the ordered sequence $0 < \alpha_0 < \alpha_1 < \ldots < \alpha_\ell < \ldots < \alpha_Z \le 1$ of α-levels for the transitive closure $\tilde{T}$ of the fuzzy tolerance T; the value α_ℓ must be found using the leap heuristic;
4. Construct the fuzzy relation $\tilde{T}_{(\alpha_\ell)}$ for the value α_ℓ;
5. Construct the initial allotment $R_I^{\alpha}(X) = \{A^l_{(\alpha_\ell)}\}$ for the fuzzy relation $\tilde{T}_{(\alpha_\ell)}$; construct the allotments which satisfy the conditions (2.4) and (2.5);
6. Construct the class of possible solutions to the classification problem $B(\alpha_\ell) = \{R^{\alpha_\ell}_{c(z)}(X)\}$ and calculate the value of the criterion $F(R^{\alpha}_{c(z)}(X), \alpha)$ for each allotment $R^{\alpha_\ell}_{c(z)}(X) \in B(\alpha_\ell)$;
7. Check the following condition:

    if for some unique allotment $R^{\alpha_\ell}_{c(z)}(X) \in B(\alpha_\ell)$ the condition (2.10) is met
    then this allotment is the classification result $R^*_c(X)$ for the value α_ℓ and stop
    else construct the set of allotments $B'(\alpha_\ell) \subseteq B(\alpha_\ell)$ which satisfy the condition (2.10) and go to step 8;
8. Perform the following operations for each allotment $R^{\alpha_\ell}_{c(z)}(X) \in B'(\alpha_\ell)$:

    8.1 Let $l := 1$;

    8.2 Find the support $Supp(A^l_{(\alpha_\ell)}) = A^l_{\alpha_\ell}$ of the fuzzy cluster $A^l_{(\alpha_\ell)} \in R^{\alpha_\ell}_{c(z)}(X)$ and construct the matrix of attributes $X_{n_l \times m_1} = [x_i^{t_1}]$, $x_i \in A^l_{\alpha_\ell}$, $t_1 = 1, \ldots, m_1$, for $A^l_{\alpha_\ell}$ where $n_l = card(A^l_{\alpha_\ell})$;

    8.3 Calculate the prototype $\tau^l = \{\bar{x}^1, \ldots, \bar{x}^{m_1}\}$ of the class $A^l_{\alpha_\ell}$ according to the formula

$$\bar{x}^{t_1} = \frac{1}{n_l} \sum_{x_i \in A^l_{\alpha_\ell}} x_i^{t_1}, \; t_1 = 1, \ldots, m_1;$$

    8.4 Calculate the distance $d(\tau^l, \tau^l)$ between the typical point τ^l of the fuzzy cluster $A^l_{(\alpha_\ell)}$ and its prototype τ^l;

    8.5 Check the following condition:

    if all fuzzy clusters $A^l_{(\alpha_\ell)} \in R^{\alpha_\ell}_{c(z)}(X)$ are not verified
    then let $l = l + 1$ and go to step 8.2
    else go to step 9

9. Compare the fuzzy clusters $A^{l}_{(\alpha_\ell)}$ which are the elements of different allotments $R^{\alpha_\ell}_{c(z)}(X) \in B'(\alpha_\ell)$; the allotment $R^{\alpha_\ell}_{c(z)}(X) \in B'(\alpha_\ell)$ for which the distance $d(\tau^l, \tau^l)$ is minimal for all fuzzy clusters $A^{l}_{(\alpha_\ell)}$ is the classification result $R^{*}_{c}(X)$.

The allotment $R^{*}_{c}(X) = \{A^{l}_{(\alpha)} \mid l = \overline{1,c}\}$ among an unknown number c of fully separated fuzzy clusters, normalized prototypes $\{\tau^1, \ldots, \tau^c\}$ of the corresponding fuzzy clusters, and the value of the tolerance threshold $\alpha \in (0,1]$ are results of the classification.

The determination of the principal allotment among an unknown number c of fully separated fuzzy clusters is the essence of the D-PAFC-TC-algorithm. The leap heuristic is not used in the D-PAFC-TC-algorithm because the principal allotment should be constructed. So, the solution of the classification problem should be obtained for the value α_1 because the condition of Lemma 1.2 is met.

The D-PAFC-TC-algorithm is a nine-step classification procedure [126].

The D-PAFC-TC-Algorithm

1. Construct the matrix of the fuzzy tolerance relation $T_{n \times n} = [\mu_T(x_i, x_j)]$ by normalizing the initial data $\hat{X}_{n \times m_1} = [\hat{x}_i^{t_1}]$, $i = 1, \ldots, n$, $t_1 = 1, \ldots, m_1$, and by choosing a suitable distance for fuzzy sets;
2. Construct the transitive closure $\tilde{T}$ of the fuzzy tolerance T according to the definition 1.24;
3. Construct the ordered sequence $0 < \alpha_0 < \alpha_1 < \ldots < \alpha_\ell < \ldots < \alpha_Z \leq 1$ of α-levels for the transitive closure $\tilde{T}$ of the fuzzy tolerance T;
4. Construct the fuzzy relation $\tilde{T}_{(\alpha_1)}$ for the value α_1;
5. Construct the initial allotment $R^{\alpha}_{I}(X) = \{A^{l}_{(\alpha_1)}\}$ for the fuzzy relation $\tilde{T}_{(\alpha_1)}$; construct the allotments which satisfy the conditions (2.4) and (2.5);
6. Construct the class of possible solutions to the classification problem $B(\alpha_1) = \{R^{\alpha_1}_{c(z)}(X)\}$ and calculate the value of the criterion $F(R^{\alpha}_{c(z)}(X), \alpha)$ for each allotment $R^{\alpha_1}_{c(z)}(X) \in B(\alpha_1)$;
7. Check the following condition:

   if for some unique allotment $R^{\alpha_1}_{c(z)}(X) \in B(\alpha_1)$ the condition (2.10) is met

   then the allotment is the classification result $R^{\alpha_1}_{P}(X)$ for the value α_1 and stop

   else construct the set of allotments $B'(\alpha_1) \subseteq B(\alpha_1)$ which satisfy the condition (2.10) and go to step 8;

8. Perform the following operations for each allotment $R^{\alpha_1}_{c(z)}(X) \in B'(\alpha_1)$:

   8.1 Let $l := 1$;

   8.2 Determine the support $Supp(A^l_{(\alpha_1)}) = A^l_{\alpha_1}$ of the fuzzy cluster $A^l_{(\alpha_1)} \in R^{\alpha_1}_{c(z)}(X)$ and construct the matrix of attributes $X_{n_l \times m_1} = [x_i^{t_1}]$, $x_i \in A^l_{\alpha_1}$, $t_1 = 1,\ldots,m_1$, for $A^l_{\alpha_1}$ where $n_l = card(A^l_{\alpha_1})$;

   8.3 Calculate the prototype $\tau^l = \{\overline{x}^1,\ldots,\overline{x}^{m_1}\}$ of the class $A^l_{\alpha_1}$ according to the formula

$$\overline{x}^{t_1} = \frac{1}{n_l} \sum_{x_i \in A^l_{\alpha_1}} x_i^{t_1} \, , \; t_1 = 1,\ldots,m_1 ;$$

   8.4 Calculate the distance $d(\tau^l, \tau^l)$ between the typical point τ^l of the fuzzy cluster $A^l_{(\alpha_1)}$ and its prototype τ^l;

   8.5 Check the following condition:
   if all fuzzy clusters $A^l_{(\alpha_1)} \in R^{\alpha_1}_{c(z)}(X)$ are not verified
   then let $l = l + 1$ and go to step 8.2
   else go to step 9

9. Compare the fuzzy clusters $A^l_{(\alpha_1)}$ which are the elements of different allotments $R^{\alpha_1}_{c(z)}(X) \in B'(\alpha_1)$, and the allotment $R^{\alpha_1}_{c(z)}(X) \in B'(\alpha_1)$ for which the distance $d(\tau^l, \tau^l)$ is minimal for all fuzzy clusters $A^l_{(\alpha_1)}$ is the classification result $R^{\alpha_1}_P(X)$.

The principal allotment $R^{\alpha_1}_P(X) = \{A^l_{(\alpha_1)} \mid l = \overline{1,c}\}$ among fully separated fuzzy clusters, the normalized prototypes $\{\tau^1,\ldots,\tau^c\}$ of the corresponding fuzzy clusters, and the value of the tolerance threshold $\alpha_1 \in (0,1]$ are results of classification.

The determination of the allotment $R^*_c(X)$ among an unknown number c of fully separated fuzzy clusters with respect to the minimal value α of the tolerance threshold is the essence of the D-AFC-TC(α)-algorithm.

So, the D-AFC-TC(α)-algorithm is a ten-step classification procedure [126].

The D-AFC-TC(α)-Algorithm

1. Construct the matrix of the fuzzy tolerance relation $T_{n \times n} = [\mu_T(x_i, x_j)]$ by normalizing the initial data $\hat{X}_{n \times m_1} = [\hat{x}_i^{t_1}]$, $i = 1,\ldots,n$, $t_1 = 1,\ldots,m_1$, and by choosing a suitable distance for the fuzzy sets;

2. Construct the transitive closure $\tilde{T}$ of the fuzzy tolerance T according to the definition 1.24;
3. Construct the ordered sequence $0 < \alpha_0 < \alpha_1 < \ldots < \alpha_\ell < \ldots < \alpha_Z \leq 1$ of α-levels for the transitive closure $\tilde{T}$ of the fuzzy tolerance T; the value α_ℓ must be found by the leap heuristic;
4. Check the following condition:

   if for the calculated value α_ℓ the condition $\alpha^* \leq \alpha_\ell$ is met

   then go to step 5

   else the value α_ℓ must be found by the leap heuristicin the ordered sequence $\alpha^* < \alpha_1 < \ldots < \alpha_\ell < \ldots < \alpha_Z \leq 1$ and go to step 5;
5. Construct the fuzzy relation $\tilde{T}_{(\alpha_\ell)}$ for the value α_ℓ;
6. Construct the initial allotment $R_I^\alpha(X) = \{A_{(\alpha_\ell)}^l\}$ for the fuzzy relation $\tilde{T}_{(\alpha_\ell)}$; construct the allotments which satisfy the conditions (2.4) and (2.5);
7. Construct the class of possible solutions of the classification problem $B(\alpha_\ell) = \{R_{c(z)}^{\alpha_\ell}(X)\}$ and calculate the value of the criterion $F(R_{c(z)}^\alpha(X), \alpha)$ for each allotment $R_{c(z)}^{\alpha_\ell}(X) \in B(\alpha_\ell)$;
8. Check the following condition:

   if for some unique allotment $R_{c(z)}^{\alpha_\ell}(X) \in B(\alpha_\ell)$ the condition (2.10) is met

   then this allotment is the classification result $R_c^*(X)$ for the value α_ℓ and stop

   else construct the set of allotments $B'(\alpha_\ell) \subseteq B(\alpha_\ell)$ which satisfy the condition (2.10) and go to step 9;
9. Perform the following operations for each allotment $R_{c(z)}^{\alpha_\ell}(X) \in B'(\alpha_\ell)$:

   9.1 Let $l := 1$;

   9.2 Find the support $Supp(A_{(\alpha_\ell)}^l) = A_{\alpha_\ell}^l$ of the fuzzy cluster $A_{(\alpha_\ell)}^l \in R_{c(z)}^{\alpha_\ell}(X)$ and construct the matrix of attributes $X_{n_l \times m_1} = [x_i^{t_1}]$, $x_i \in A_{\alpha_\ell}^l$, $t_1 = 1, \ldots, m_1$, for $A_{\alpha_\ell}^l$ where $n_l = card(A_{\alpha_\ell}^l)$;

   9.3 Calculate the prototype $\tau^l = \{\bar{x}^1, \ldots, \bar{x}^{m_1}\}$ of the class $A_{\alpha_\ell}^l$ according to thd formula

$$\bar{x}^{t_1} = \frac{1}{n_l} \sum_{x_i \in A_{\alpha_\ell}^l} x_i^{t_1}, \; t_1 = 1, \ldots, m_1;$$

   9.4 Calculate the distance $d(\tau^l, \tau^l)$ between the typical point τ^l of the fuzzy cluster $A_{(\alpha_\ell)}^l$ and its prototype τ^l;

9.5 Check the following condition:
if all fuzzy clusters $A^{l}_{(\alpha_\ell)} \in R^{\alpha_\ell}_{c(z)}(X)$ are not verified
then let $l = l + 1$ and go to step 9.2
else go to step 10

10. Compare the fuzzy clusters $A^{l}_{(\alpha_\ell)}$ which are elements of different allotments $R^{\alpha_\ell}_{c(z)}(X) \in B'(\alpha_\ell)$, and the allotment $R^{\alpha_\ell}_{c(z)}(X) \in B'(\alpha_\ell)$ for which the distance $d(\tau^{l}, \tau^{l})$ is minimal for all fuzzy clusters $A^{l}_{(\alpha_\ell)}$ is the classification result $R^{*}_{c}(X)$.

Thus, the allotment $R^{*}_{c}(X) = \{A^{l}_{(\alpha)} \mid l = \overline{1,c}\}$ among an unknown number c of fully separated fuzzy clusters, the normalized prototypes $\{\tau^{1},...,\tau^{c}\}$ of the corresponding fuzzy clusters, and the value of the tolerance threshold $\alpha \geq \alpha^{*}$, $\alpha \in (0,1]$, are the results of classification. Notice that the minimal value of the tolerance threshold α^{*} is the parameter of the D-AFC-TC(α)-algorithm and this value should be estimated by an expert.

2.3.2 An Algorithm of Hierarchical Clustering

It is well-known that the transitive closure $\tilde{T}$ of a fuzzy tolerance relation T generates a hierarchical classification [60], [106].

Let us consider a hierarchical clustering procedure which was presented in [119]. The clustering procedure was called the H-AFC-TC-algorithm. Its essence is the construction of a dendrogram. We call a dendrogram a tree the leaves of which are samples subjected to classification.

Some concepts should first be defined.

Definition 2.6. *Let $0 < \alpha_0 < \alpha_1 < \ldots < \alpha_\ell < \ldots < \alpha_Z \leq 1$ be an ordered sequence of values of the tolerance threshold. The allotment $R^{\alpha_\ell}_{c(z)}(X)$ of the set of objects among anunknown number c of fully separatde fuzzy clusters for some tolerance threshold $\alpha_\ell \in (0,1]$ is the allotment of the set $X = \{x_1,...,x_n\}$ for the corresponding value of $\alpha_\ell \in (0,1]$.*

Thus, the class of possible solutions to the classification problem $B(\alpha_\ell) = \{R^{\alpha_\ell}_{c(z)}(X)\}$ can be constructed for the corresponding value α_ℓ and the allotment $R^{*}_{\alpha_\ell}(X)$ is the classification result for the value α_ℓ.

Definition 2.7. *The allotment* $R_T^{\alpha_\ell}(X)$ *of the set of objects among the unique fuzzy cluster* $A^1_{(\alpha_\ell)}$ *for some tolerance threshold* $\alpha_\ell \in (0,1]$ *is the total allotment of the set* $X = \{x_1,\ldots,x_n\}$.

It is should be noted that the total allotment $R_T^{\alpha_\ell}(X)$ can be obtained for the value α_0 from the ordered sequence $0 < \alpha_0 < \alpha_1 < \ldots < \alpha_\ell < \ldots < \alpha_Z \le 1$ because the condition of lemma 1.2 is met.

Definition 2.8. *Let* $0 < \alpha_0 < \alpha_1 < \ldots < \alpha_\ell < \ldots < \alpha_Z \le 1$ *be the ordered sequence ofvalues of the tolerance threshold and* $\{R^*_{\alpha_0}(X),\ldots,R^*_{\alpha_Z}(X)\}$ *be the corresponding sequence of allotments. Let be* $A^l_{(\alpha_\ell)}$ *and* $A^l_{(\alpha_{\ell+1})}$ *be two fuzzy clusters from the corresponding different allotments* $R^*_{\alpha_\ell}(X)$ *and* $R^*_{\alpha_{\ell+1}}(X)$. *If the condition* $Supp\left(A^l_{(\alpha_{\ell+1})}\right) \subseteq Supp\left(A^l_{(\alpha_\ell)}\right)$ *is met for all fuzzy clusters and all values* $\ell \in \{0,\ldots,Z\}$, *then the allotment among the fuzzy clusters* $R^*_{\alpha_{\ell+1}}(X)$ *is included in the allotment among fuzzy clusters* $R^*_{\alpha_\ell}(X)$, $R^*_{\alpha_{\ell+1}}(X) \subseteq R^*_{\alpha_\ell}(X)$.

In other words, if $R^*_{\alpha_\ell}(X) = \{A^1_{(\alpha_\ell)},\ldots,A^{c_\ell}_{(\alpha_\ell)}\}$ and $R^*_{\alpha_{\ell+1}}(X) = \{A^1_{(\alpha_{\ell+1})},\ldots,A^{c_{\ell+1}}_{(\alpha_{\ell+1})}\}$ are two allotments among an unknown number of fully separated fuzzy clusters for the corresponding values of the tolerance threshold $\alpha_\ell, \alpha_{\ell+1} \in (0,1]$, and the support of each element of the allotment $R^*_{\alpha_\ell}(X)$ can be expressed the union of supports of several elements of the allotment $R^*_{\alpha_{\ell+1}}(X)$ then the allotment $R^*_{\alpha_{\ell+1}}(X)$ is said to be a refinement of the allotment $R^*_{\alpha_\ell}(X)$.

Definition 2.9. *Let* $0 < \alpha_0 < \alpha_1 < \ldots < \alpha_\ell < \ldots < \alpha_Z \le 1$ *be the ordered sequence ofvalues of the tolerance threshold. A hierarchy of allotments is the ordered sequence of allotments* $R^*_{\alpha_0}(X) \subseteq R^*_{\alpha_1}(X) \subseteq \ldots R^*_{\alpha_\ell}(X) \subseteq \ldots$.

It is should be noted, that the value $\ell \in \{0,\ldots,Z\}$ can be interpreted as a stratification index of the hierarchy of allotments. If the hierarchy of allotments is constructed for all values $\ell = 0,\ldots,Z$, then the hierarchy is called the full hierarchy.

The construction of the full hierarchy of allotments $R^*_{\alpha_0}(X) \subseteq R^*_{\alpha_1}(X) \subseteq \ldots R^*_{\alpha_\ell}(X) \subseteq \ldots \subseteq R^*_{\alpha_Z}(X)$ is the essence of the H-AFC-TC-algorithm.

The H-AFC-TC-algorithm is an eleven-step procedure of classification [119].

The H-AFC-TC-Algorithm

1. Construct the matrix of the fuzzy tolerance relation $T_{n \times n} = [\mu_T(x_i, x_j)]$ by normalizing the initial data $\hat{X}_{n \times m_1} = [\hat{x}_i^{t_1}]$, $i = 1, \ldots, n$, $t_1 = 1, \ldots, m_1$, and by choosing a suitable distance for fuzzy sets;
2. Construct the transitive closure $\tilde{T}$ of the fuzzy tolerance T according to the definition 1.24;
3. Construct the ordered sequence $0 < \alpha_0 < \alpha_1 < \ldots < \alpha_\ell < \ldots < \alpha_Z \leq 1$ of α-levels for the transitive closure $\tilde{T}$ of the fuzzy tolerance T;
4. Let $\ell := 0$;
5. Construct the fuzzy relation $\tilde{T}_{(\alpha_\ell)}$ for the value α_ℓ;
6. Construct the initial allotment $R_I^{\alpha_\ell}(X) = \{A^l_{(\alpha_\ell)}\}$ for the fuzzy relation $\tilde{T}_{(\alpha_\ell)}$; construct the allotments which satisfy the conditions (2.4) and (2.5);
7. Construct the class of possible solutions of the classification problem $B(\alpha_\ell) = \{R^{\alpha_\ell}_{c(z)}(X)\}$ and calculate the value of the criterion $F(R^{\alpha}_{c(z)}(X), \alpha)$ for each allotment $R^{\alpha_\ell}_{c(z)}(X) \in B(\alpha_\ell)$;
8. Check the following condition:

   if for some unique allotment $R^{\alpha_\ell}_{c(z)}(X) \in B(\alpha_\ell)$ the condition (2.10) is met

   then the allotment is the classification result $R^*_{\alpha_\ell}(X)$ for the value α_ℓ and stop

   else construct the set of allotments $B'(\alpha_\ell) \subseteq B(\alpha_\ell)$ which satisfy the condition (2.10) and go to step 9;
9. Perform the following operations for each allotment $R^{\alpha_\ell}_{c(z)}(X) \in B'(\alpha_\ell)$:

   9.1 Let $l := 1$;

   9.2 Detect the support $Supp(A^l_{(\alpha_\ell)}) = A^l_{\alpha_\ell}$ of the fuzzy cluster $A^l_{(\alpha_\ell)} \in R^{\alpha_\ell}_{c(z)}(X)$ and construct the matrix of attributes $X_{n_l \times m_1} = [x_i^{t_1}]$, $x_i \in A^l_{\alpha_\ell}$, $t_1 = 1, \ldots, m_1$, for $A^l_{\alpha_\ell}$ where $n_l = card(A^l_{\alpha_\ell})$;

   9.3 Calculate the prototype $\tau^l = \{\bar{x}^1, \ldots, \bar{x}^{m_1}\}$ of the class $A^l_{\alpha_\ell}$ according to the formula

$$\bar{x}^{t_1} = \frac{1}{n_l} \sum_{x_i \in A^l_{\alpha_\ell}} x_i^{t_1}, \; t_1 = 1, \ldots, m_1;$$

9.4 Calculate the distance $d(\tau^l, \tau^l)$ between the typical point τ^l of the fuzzy cluster $A^l_{(\alpha_\ell)}$ and its prototype τ^l;

9.5 Check the following condition:

if all fuzzy clusters $A^l_{(\alpha_\ell)} \in R^{\alpha_\ell}_{c(z)}(X)$ are not verified
then let $l = l + 1$ and go to step 9.2
else go to step 10

10. Compare the fuzzy clusters $A^l_{(\alpha_\ell)}$ which are the elements of different allotments $R^{\alpha_\ell}_{c(z)}(X) \in B'(\alpha_\ell)$, and the allotment $R^{\alpha_\ell}_{c(z)}(X) \in B'(\alpha_\ell)$ for which the distance $d(\tau^l, \tau^l)$ is minimal for all fuzzy clusters $A^l_{(\alpha_\ell)}$ is the classification result $R^*_{\alpha_\ell}(X)$ for the value α_ℓ;
11. Check the following condition:

if the condition $\ell \leq Z$ is met
then let $\ell := \ell + 1$ and go to step 5;
else stop.

It should be noted that the H-AFC-TC-algorithm is a divisive clustering procedure. The full hierarchy of allotments $R^*_{\alpha_0}(X) \subseteq R^*_{\alpha_1}(X) \subseteq \ldots R^*_{\alpha_\ell}(X) \subseteq \ldots \subseteq R^*_{\alpha_Z}(X)$ is the main result of classification. For lack of space, details on the construction of a dendrogram are omitted as they can easily be found.

2.3.3 Numerical Examples

The formula (2.11) and the squared normalized Euclidean distance (2.15) were used in the experiments with the direct prototype-based clustering procedures. First, let us consider results of the direct prototype-based algorithms applied to Yang and Wu's [168] two-dimensional data set.

By executing the D-AFC-TC-algorithm, we obtain the allotment $R^*_c(X)$ among $c = 4$ fully separated fuzzy clusters, which corresponds to the result, is received for the tolerance threshold $\alpha = 0.9994$. Membership functions of four classes are presented in Figure 2.32. The membership values of the first class are represented by ○, the membership values of the second class are represented by ■, the membership values of the third class are represented by ▲, and the membership values of the fourth class are represented by □.

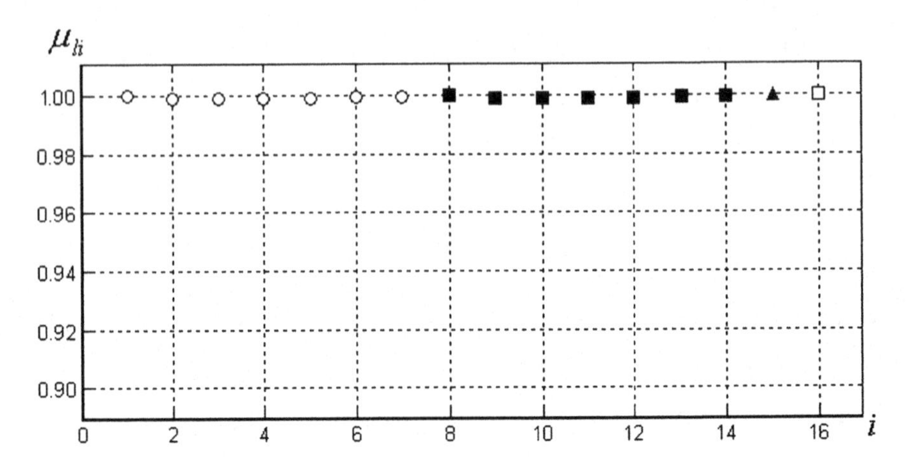

Fig. 2.32 Membership functions of four fuzzy clusters obtained from the D-AFC-TC-algorithm

The value of the membership function of the fuzzy cluster which corresponds to the first class is maximal for the first object and is equal one. So, the first object is the typical point of the first fuzzy cluster. The membership value of the eighth object is equal one for the second fuzzy cluster. Thus, the eighth object is the typical point of the second fuzzy cluster. The membership value of the fifteenth object is equal one for the third fuzzy cluster. That is why the fifteenth object is the typical point of the third fuzzy cluster. The value of the membership function of the fuzzy cluster which corresponds to the fourth class is maximal for the sixteenth object and this object is the typical point of the corresponding cluster. Notice that the fifteenth object is a bridge object and the sixteenth object is an outlier. Both objects are unique elements of the corresponding fuzzy clusters.

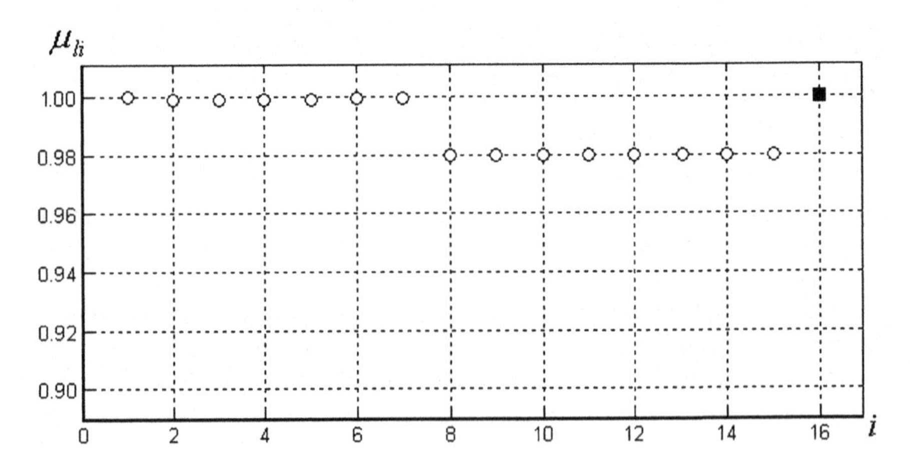

Fig. 2.33 Membership functions of two fuzzy clusters obtained from the D-PAFC-TC-algorithm

By executing the D-PAFC-TC-algorithm, the principal allotment $R_P^{0.98}(X)$ among two fuzzy clusters, which corresponds to the result, is obtained for the tolerance threshold $\alpha = 0.98$. The membership functions of two classes are presented in Figure 2.33 where the membership values of the first class are represented by ○ and the membership values of the second class are represented by ■.

Thus, the first object is the typical point of the first fuzzy cluster and the sixteenth object is the typical point of the fuzzy cluster which corresponds to the second class. It should be noted that the sixteenth object is an outlier and this object is the unique element of the second fuzzy cluster.

Second, let us consider results of the direct prototype-based algorithms applied to Sneath and Sokal's [104] two-dimensional data set.

By executing the D-AFC-TC-algorithm, we obtain the allotment $R_c^*(X)$ among $c = 2$ fully separated fuzzy clusters which corresponds to the result, obtained for the tolerance threshold $\alpha = 0.96875$. The value of the membership function of the fuzzy cluster which corresponds to the first class is maximal for the sixth object and is equal one. So, the sixth object is the typical point of the first fuzzy cluster. The value of the membership function of the fuzzy cluster which corresponds to the second class is maximal for the tenth object and the object is the typical point of the corresponding cluster.

On the other hand, by executing the D-PAFC-TC-algorithm, we obtain the results which are equal to the results obtained from the D-AFC-TC-algorithm.

The membership functions of these two fuzzy clusters are presented in Figure 2.34. The membership values of the first class are represented by ○ and the membership values of the second class are represented by ■.

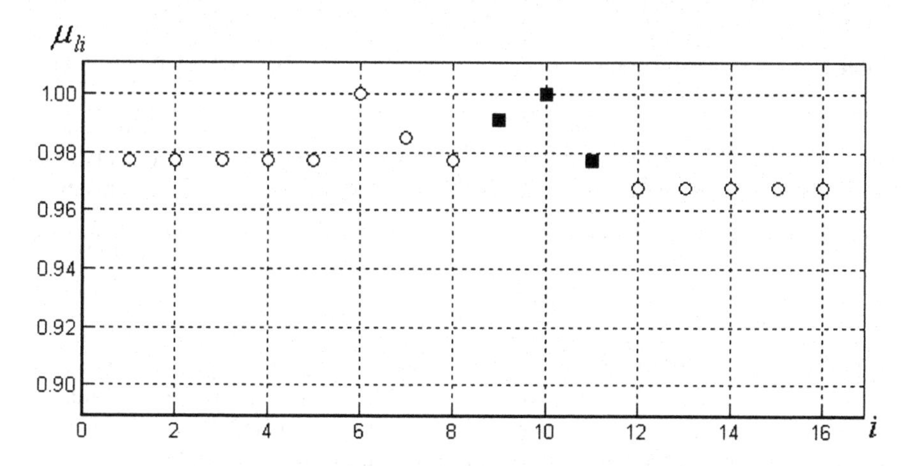

Fig. 2.34 Membership functions of two fuzzy clusters obtained from the D-AFC-TC-algorithm

Third, let us consider the results of the D-AFC-TC-algorithm applied to Anderson's Iris data set [4]. The value of accuracy threshold was selected as $\varepsilon = 0.0001$ in the experiment. By executing the D-AFC-TC-algorithm, we obtain the following: the first class is formed by 50 elements all being the Iris Setosa and the second class is formed by 100 elements all being the Iris Versicolor and Iris Virginica. The allotment $R_c^*(X)$ among $c = 2$ fully separated fuzzy clusters, which corresponds to this result, is obtained for the tolerance threshold $\alpha = 0.9929$. The object x_{95} and the object x_{106} are typical points of the first fuzzy cluster. The object x_3 is the typical point of the fuzzy cluster which corresponds to the second class. The membership functions of two classes are presented in Figure 2.35. The membership values of the first fuzzy cluster are represented by + and the membership values of the second fuzzy cluster are represented by ■.

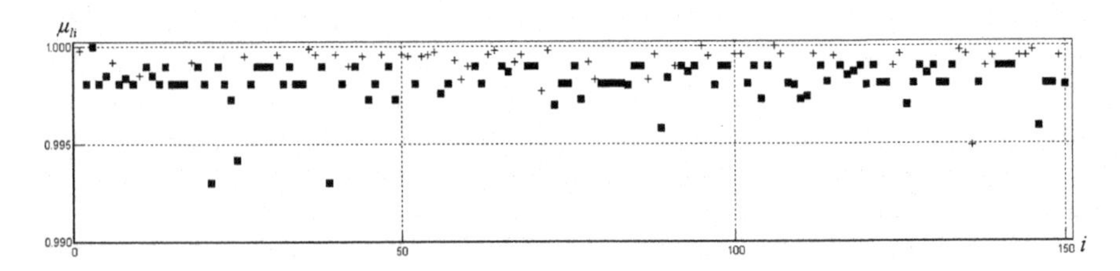

Fig. 2.35 Membership functions of two clusters obtained from the D-AFC-TC-algorithm

The main characteristic features of the classification result obtained by using the D-AFC-TC-algorithm are presented in Table 2.12.

Table 2.12 Characteristic features of the result obtained by using the D-AFC-TC-algorithm

Number of classes	Value of the modified linear index of fuzziness (2.21)	Value of the modified quadratic index of fuzziness (2.24)	Value of the density of fuzzy cluster (2.28)
1	0.0014	0.0021	0.9993
2	0.0036	0.0042	0.9982

Let us consider briefly results of the H-AFC-TC-algorithm applied to the two two-dimensional data sets. The formula (2.11) and the normalized Euclidean distance (2.14) are used in both experiments. For Yang and Wu's [167] two-dimensional data set, we obtain the dendrogram shown in Figure 2.36.

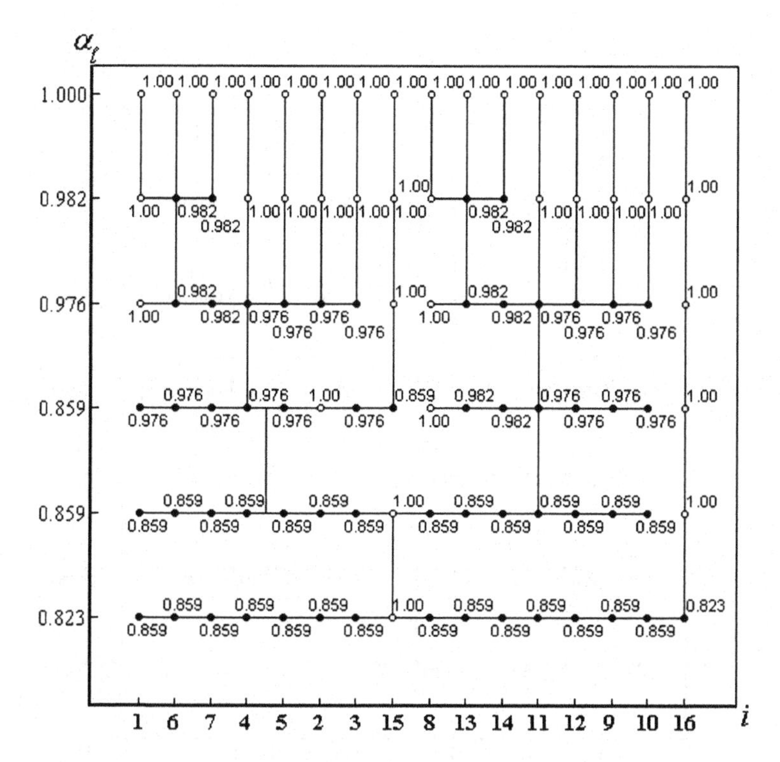

Fig. 2.36 The dendrogram for Yang and Wu's data set

It should be noted that the values α_1 and α_2 are equal 0.859 in Figure 2.36. The value of the accuracy threshold has been selected as $\varepsilon = 0.001$. So, the condition $\alpha_1 < \alpha_2$ is met.

The dendrogram obtained from the H-AFC-TC-algorithm for Sneath and Sokal's two-dimensional data set is presented in Figure 2.37.

Typical points of the fuzzy clusters are represented by ○ in both figures. So, the results obtained with the H-AFC-TC-algorithm for Yang and Wu's data set and Sneath and Sokal's data set illustrate the effectiveness of the H-AFC-TC-algorithm for constructing the hierarchy of allotments.

All prototype-based clustering procedures can be applied directly to the data given as the two-way data matrix (1.69). The results of application of the proposed algorithms to well-known data sets indicate that these algorithms can be very useful in the exploratory data analysis.

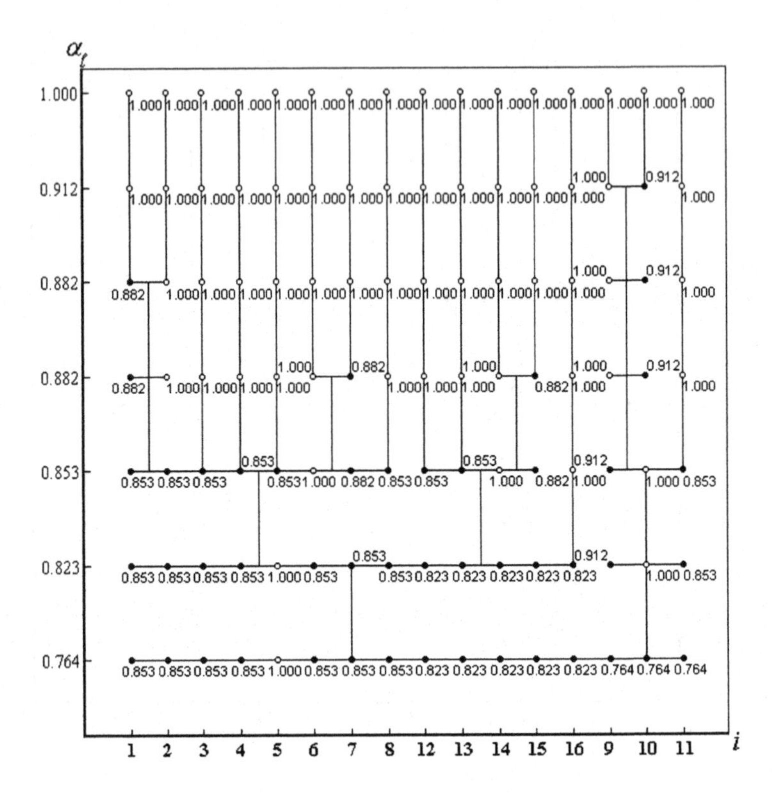

Fig. 2.37 The dendrogram for Sneath and Sokal's data set

Chapter 3
Clustering Approaches for Uncertain Data

Most fuzzy clustering techniques are designed for handling crisp data. However, data can clearly be uncertain. In other words, there exists a feature of the object that may assume several values at the same time or, for a given feature, there exists uncertainty as to its values. Traditional fuzzy clustering methods cannot be applied directly to such types of objects. So, a problem of fuzzy clustering of uncertain data arises. It is of relevance in many areas, notably in military applicaions, medicine, biology, chemistry, economy, sociology and many other domains.

Several kinds of uncertainty exist and anumber of approaches to the uncertain data fuzzy clustering problem solving have been proposed by different researchers. A brief review of uncertain data clustering methods is given by Kreinovich and Kosheleva [65]. In particular, they distinguish the uncertainty of the input data and the uncertainty of the output data. Fuzziness and equivocation can be considerd as the basic types of uncertainty of the output data. Fuzzy clustering methods can be useful in the case of fuzziness of the output data and the possibilistic approach to clustering seems to be appropriate in the case of equivocation of the output data.

The uncertainty of the input data can be divided into three types: interval uncertainty, probabilistic uncertainty and fuzzy uncertainty. It should be noted that the interval-valued data is a particular case of the three-way data. Corresponding clustering techniques should be used in these cases. Moreover, a detailed discussion of the problem of uncertainty in cluster analysis is also presented in [113]. For example, ambiguity can be considered as a special case of uncertainty in clustering problems. Degrees of expressiveness of attribute for each object and degrees of their non-expressiveness may be presented in the matrix of initial data. Values of the degrees of similarity and a difference between objects can be represented in the matrix of relational data simultaneously. The ambiguity of the output data can be expressed at the same time by the membership and non-membership values in matrix of the fuzzy partition. That is why the methods of intuitionistic fuzzy clustering seem to be appropriate in the case of ambiguous input data as well as in the case of ambiguity of the output clustering structure.

Therefore, the ambiguous data, the fuzzy data, the three-way data and the interval-value data can be considered as basic kinds of uncertain data. The applications of the proposed clustering approach to the analysis of uncertain data of the above kinds are considered in this present chapter.

D.A. Viattchenin: *A Heuristic Approach to Possibilistic Clustering*, Studfuzz 297, pp. 119–182.
DOI: 10.1007/978-3-642-35536-3_3

3.1 A Clustering Approach Based on Intuitionistic Fuzzy Relations

In the first subsection of this section, a brief review of intuitionistic fuzzy clustering approaches is provided, in the second subsection some basic concepts of the Atanassov's intuitionistic fuzzy set theory are considered, in the third subsection a method of the decomposition of intuitionistic fuzzy tolerances is presented, in the fourth subsection basic definitions of the intuitionistic generalization of the clustering method are introduced, in the fifth subsection an outline of the D-PAIFC-algorithm is described, the sixth subsection includes the data description for two Hung's numerical examples and results of their processing by the proposed D-PAIFC-algorithm in comparison with Hung's clustering algorithm.

3.1.1 A Brief Review of the Existing Intuitionistic Fuzzy Clustering Techniques

The intuitionistic fuzzy set theory, proposed by Atanassov [4], [5], has been used in a wide range of applications, such as data analysis, logic programming, medical diagnosis, and decision making. Applications of the intuitionistic fuzzy set theory to pattern recognition problems are outlined by Vlachos and Sergiadis[150]. Moreover, Szmidt and Kacprzyk [105] proposed an effective and efficient approach to the classification of nominal data using the intuitionistic fuzzy sets. However, so far there has been little research effert on the clustering techniques based on the intuitionistic fuzzy sets.

A few intuitionistic fuzzy clustering methods have been proposed by different researchers. Similarly as in the case of all fuzzy clustering algorithms, object and relational procedures can be distinguished as the two basic types of the intuitionistic fuzzy clustering methods. Let us consider the relational algorithms of intuitionistic fuzzy clustering.

First, a fuzzy clustering method based on intuitionistic fuzzy tolerance relations was proposed by Hung, Lee and Fuh [52]. An intuitionistic fuzzy similarity relation matrix is obtained by starting with an intuitionistic fuzzy tolerance relation matrix using the extended n-step procedure by using the composition of intuitionistic fuzzy relations. A hard partition for the corresponding thresholds values α and β is the result of classification. Several types of the $\max{-}\mathrm{T}\,\&\,\min{-}\mathrm{S}$ compositions can be used in Hung, Lee and Fuh's [52] approach where T is some T-norm (1.10) – (1.12) and S is a corresponding S-norm (1.17) – (1.19). So, Hung, Lee and Fuh's [52] clustering method can be considered as an extended version of the clustering technique which was proposed by Yang and Shih [166].

Second, the concepts of an association matrix and an equivalent association matrix were defined by Xu, Chen and Wu [160], and methods for calculating the association coefficients of the intuitionistic fuzzy sets were introduced there.

The clustering algorithm proposed in [160] uses association coefficients of intuitionistic fuzzy sets to construct an association matrix, and employs a procedure to transform it into an equivalent association matrix. The α-cutting matrix of the equivalent association matrix is used to cluster the given intuitionistic fuzzy sets. So, a hard partition for some value of α is the result of classification. That is why the clustering method proposed in [160] is similar to the clustering technique which was proposed by Hung, Lee and Fuh [52].

Third, Cai, Lei and Zhao [17] presented a clustering technique based on the use of the intuitionistic fuzzy dissimilarity matrix and (α, β)-cutting matrices. The method is based on the transitive closure based technique. Thus, the method is also similar to that described in [52].

Fourth, a method for constructing an intuitionistic fuzzy tolerance matrix from a set of intuitionistic fuzzy sets and a netting method to clustering of intuitionistic fuzzy sets via the corresponding intuitionistic fuzzy tolerance matrix was developed by Wang, Xu, Liu and Tang [153]. A hard partition is the result of classification and the clustering result depends on a chosen value of the confidence level $\alpha \in [0,1]$. Notice that all relational algorithms cannot provide information about the membership degrees of objects to each cluster.

Let us consider the intuitionistic fuzzy clustering methods which are based on the representation of the initial data by a matrix of attributes. Some of them are objective function-based clustering procedures.

First, Pelekis, Iakovidis, Kotsifakos and Kopanakis [93] proposed a variant of the well-known FCM-algorithm that copes with uncertainty and involves a similarity measure between the intuitionistic fuzzy sets which is appropriately integrated into the clustering algorithm. The ordinary fuzzy c-partition is the clustering result. An application of the proposed clustering technique to image segmentation was described in [54].

Second, Torra, Miyamoto, Endo and Domingo-Ferrer [108] proposed a clustering method, based on the FCM-algorithm, to construct an intuitionistic fuzzy partition. In this clustering method, the intuitionistic fuzzy partition deals with the uncertainty present in different executions of the same clustering procedure. Intuitionistic fuzzy partitions for the traditional fuzzy c-means, intuitionistic fuzzy partitions for the entropy-based fuzzy c-means, and intuitionistic fuzzy partitions for the fuzzy c-means with tolerance are considered in [108].

Third, an intuitionistic fuzzy approach to distributed fuzzy clustering (IFDFC, for short) is considered by Karthikeyani, Visalakshi, Thangaveland Parvathi [59]. The corresponding IFDFC-algorithm is carried out on two different levels: a local level and a global level. On the local level, ordinary numerical data are converted into the intuitionistic fuzzy data and these data are clustered independently from each other using a modified FCM-algorithm. On the global level, a global centroid is calculated by clustering all local cluster centroids and this global centroid is again transmitted to local sites to update the local cluster model.

Fourth, a simple clustering technique based on the calculation of clusteretalonswas proposed by Todorova and Vassilev [107]. Their clustering technique was based on the assumption that the number of clusters is equal two. The iterative algorithm is stopped when all objects are assigned to crisp clusters according to a preliminarily chosen similarity measure.

Fifth, agglomerative hierarchical clustering algorithms for the classification of ordinary intuitionistic fuzzy sets and interval-valued intuitionistic fuzzy sets were proposed by Xu [161]. In each stage of both binary clustering procedures, the center of each cluster should be recalculated by using the average of the intuitionistic fuzzy sets assigned to the cluster, and the distance between two clusters should be determined as the distance between the centers of each cluster.

Sixth, an intuitionistic fuzzy c-means method (IFCM, for short) to the clustering of intuitionistic fuzzy sets was developed by Xu and Wu [162]. The corresponding IFCM-algorithm assumed that the initial data are represented as a set of intuitionistic fuzzy sets defined on the universe of attributes. The method was extended to the clustering of interval-valued intuitionistic fuzzy sets and the corresponding IVIFCM-algorithm was described in [162]. The fuzzy c-partition in the sense of (1.71) is the clustering result obtained from both algorithms.

3.1.2 Basic Concepts of the Intuitionistic Fuzzy Set Theory

Let us remind some basic definitions of the Atanassov's intuitionistic fuzzy set theory. All concepts will be considered for a finite universe $X = \{x_1, \ldots, x_n\}$.

Definition 3.1. *An intuitionistic fuzzy set* IA *in* X *is given by the ordered triple* $IA = \{\langle x_i, \mu_{IA}(x_i), \nu_{IA}(x_i)\rangle \mid x_i \in X\}$, *where* $\mu_{IA}, \nu_{IA} : X \to [0,1]$ *satisfy the condition*

$$0 \le \mu_{IA}(x_i) + \nu_{IA}(x_i) \le 1, \tag{3.1}$$

for all $x_i \in X$. *The values* $\mu_{IA}(x_i)$ *and* $\nu_{IA}(x_i)$ *denote the degree of membership and the degree of non-membership of element* $x_i \in X$ *to* IA, *respectively.*

For each intuitionistic fuzzy set IA in X, an intuitionistic fuzzy index [4] of an element $x_i \in X$ in IA can be defined as follows

$$\rho_{IA}(x_i) = 1 - (\mu_{IA}(x_i) + \nu_{IA}(x_i)). \tag{3.2}$$

The intuitionistic fuzzy index $\rho_{IA}(x_i)$ can be considered as a hesitancy degree of x_i to IA. Clearly, $0 \le \rho_{IA}(x_i) \le 1$, for all $x_i \in X$.

Obviously, when $\nu_{IA}(x_i) = 1 - \mu_{IA}(x_i)$, for every $x_i \in X$, the intuitionistic fuzzy set IA boils down to the ordinary fuzzy set A in X. For each fuzzy set A in X, we have $\rho_A(x_i) = 0$, for all $x_i \in X$.

Let $\mathrm{IFS}(X)$ denote the set of all intuitionistic fuzzy sets in X. The basic operations on te intuitionistic fuzzy sets were defined by Atanassov [4], [5] and in his other publications. In particular, if $IA, IB \in \mathrm{IFS}(X)$ then

$$IA \cap IB = \{\langle x_i, \min(\mu_{IA}(x_i), \mu_{IB}(x_i)), \max(\nu_{IA}(x_i), \nu_{IB}(x_i))\rangle \mid x_i \in X\}, \quad (3.3)$$

and

$$IA \cup IB = \{\langle x_i, \max(\mu_{IA}(x_i), \mu_{IB}(x_i)), \min(\nu_{IA}(x_i), \nu_{IB}(x_i))\rangle \mid x_i \in X\}. \quad (3.4)$$

Moreover, some properties of the intuitionistic fuzzy sets were given in [14]. For example, if $IA, IB \in \mathrm{IFS}(X)$, then

$$IA \leq IB \Leftrightarrow \mu_{IA}(x_i) \leq \mu_{IB}(x_i) \text{ and } \nu_{IA}(x_i) \geq \nu_{IB}(x_i), \forall x_i \in X, \quad (3.5)$$

$$IA \preceq IB \Leftrightarrow \mu_{IA}(x_i) \leq \mu_{IB}(x_i) \text{ and } \nu_{IA}(x_i) \leq \nu_{IB}(x_i), \forall x_i \in X, \quad (3.6)$$

$$IA = IB \Leftrightarrow IA \leq IB \text{ and } IA \geq IB, \forall x_i \in X, \quad (3.7)$$

$$\overline{IA} = \{\langle x_i, \nu_{IA}(x_i), \mu_{IA}(x_i)\rangle \mid x_i \in X\}. \quad (3.8)$$

Let us consider some definitions which will be needed forour further considerations. In particular, an α, β-level set of an intuitionistic fuzzy set IA in X must be introduced.

Definition 3.2. *The an α, β-level set of an intuitionistic fuzzy set IA in X can be defined as*

$$IA_{\alpha,\beta} = \{x_i \in X \mid \mu_{IA}(x_i) \geq \alpha, \nu_{IA}(x_i) \leq \beta\}, \quad (3.9)$$

where the condition

$$0 \leq \alpha + \beta \leq 1, \quad (3.10)$$

is satisfied for any values α and β, $\alpha, \beta \in [0,1]$.

The concept of the α-level fuzzy set can be extended for the intuitionistic fuzzy sets. So, a concept of the (α, β)-level intuitionistic fuzzy set was introduced in [130] as follows.

Definition 3.3. *The (α, β)-level intuitionistic fuzzy set $IA_{(\alpha,\beta)}$ in X is given by the following expression:*

$$IA_{(\alpha,\beta)} = \left\{ \left\langle x_i \in IA_{\alpha,\beta}, \mu_{IA_{(\alpha,\beta)}}(x_i) = \mu_{IA}(x_i), \nu_{IA_{(\alpha,\beta)}}(x_i) = \nu_{IA}(x_i) \right\rangle \right\}, \quad (3.11)$$

where $\alpha, \beta \in [0,1]$ satisfies the condition (3.10), and $IA_{\alpha,\beta}$ is the α, β -level set of an intuitionistic fuzzy set IA defined by (3.9).

In other words, if IA is an intuitionistic fuzzy set in X, where X is the set of elements, then the (α, β) -level intuitionistic fuzzy set $IA_{(\alpha,\beta)}$ in X, for which

$$\mu_{IA_{(\alpha,\beta)}}(x_i) = \begin{cases} \mu_{IA}(x_i), if \ \mu_{IA}(x_i) \geq \alpha \\ 0, \quad otherwise \end{cases}, \quad (3.12)$$

and

$$\nu_{IA_{(\alpha,\beta)}}(x_i) = \begin{cases} \nu_{IA}(x_i), if \ \nu_{IA}(x_i) \leq \beta \\ 0, \quad otherwise \end{cases}. \quad (3.13)$$

is called an (α, β) -level intuitionistic fuzzy subset $IA_{(\alpha,\beta)}$ of the intuitionistic fuzzy set IA in X for some $\alpha, \beta \in [0,1]$, $0 \leq \alpha + \beta \leq 1$. Obviously, the condition $IA_{(\alpha,\beta)} \preceq IA$ is met for any intuitionistic fuzzy set IA and its (α, β) - level intuitionistic fuzzy subset $IA_{(\alpha,\beta)}$, for any $\alpha, \beta \in [0,1]$, $0 \leq \alpha + \beta \leq 1$. This very important property will be useful in further considerations.

Let us remind some definitions which were discussed by Burillo and Bustince [14], [15].

Definition 3.4. *Let $X = \{x_1, \ldots, x_n\}$ be an ordinary non-empty set. The binary intuitionistic fuzzy relation IR on X is an intuitionistic fuzzy subset IR of $X \times X$, which is given by the expression*

$$IR = \left\{ \left\langle (x_i, x_j), \mu_{IR}(x_i, x_j), \nu_{IR}(x_i, x_j) \right\rangle \mid x_i, x_j \in X \right\}, \quad (3.14)$$

where $\mu_{IR} : X \times X \to [0,1]$ and $\nu_{IR} : X \times X \to [0,1]$ satisfy the condition $0 \leq \mu_{IR}(x_i, x_j) + \nu_{IR}(x_i, x_j) \leq 1$, for each $(x_i, x_j) \in X \times X$.

Let $\mathrm{IFR}(X)$ denote the set of all intuitionistic fuzzy relations on some universe X. Let us consider basic properties of the intuitionistic fuzzy relations. Let A, B be some T-norms given by (1.10) – (1.12) and Λ, P be some S-norms given by (1.17) – (1.19), and $IR, IQ \in \mathrm{IFR}(X)$ be two binary intuitionistic fuzzy relations on X.

Definition 3.5. *The composed relation* $IR \underset{\Lambda,\mathrm{P}}{\overset{\mathrm{A,B}}{\circ}} IQ \in \mathrm{IFR}(X)$ *is defined by*

$$IR \underset{\Lambda,\mathrm{P}}{\overset{\mathrm{A,B}}{\circ}} IQ = \left\{ \left\langle (x_i, x_k), \mu_{IR \underset{\Lambda,\mathrm{P}}{\overset{\mathrm{A,B}}{\circ}} IQ}(x_i, x_k), \nu_{IR \underset{\Lambda,\mathrm{P}}{\overset{\mathrm{A,B}}{\circ}} IQ}(x_i, x_k) \right\rangle \middle| x_i, x_k \in X \right\}, \quad (3.15)$$

where

$$\mu_{IR \underset{\Lambda,\mathrm{P}}{\overset{\mathrm{A,B}}{\circ}} IQ}(x_i, x_k) = \underset{x_j}{\mathrm{A}}\{\mathrm{B}[\mu_{IR}(x_i, x_j), \mu_{IQ}(x_j, x_k)]\}, \quad (3.16)$$

and

$$\nu_{IR \underset{\Lambda,\mathrm{P}}{\overset{\mathrm{A,B}}{\circ}} IQ}(x_i, x_k) = \underset{x_j}{\Lambda}\{\mathrm{P}[\nu_{IR}(x_i, x_j), \nu_{IQ}(x_j, x_k)]\}. \quad (3.17)$$

Clearly, the following condition

$$0 \le \mu_{IR \underset{\Lambda,\mathrm{P}}{\overset{\mathrm{A,B}}{\circ}} IQ}(x_i, x_k) + \nu_{IR \underset{\Lambda,\mathrm{P}}{\overset{\mathrm{A,B}}{\circ}} IQ}(x_i, x_k) \le 1, \quad (3.18)$$

must be met for all $(x_i, x_k) \in X \times X$ in the previous definition.

An intuitionistic fuzzy relation $IR \in \mathrm{IFR}(X)$ is reflexive if for each $x_i \in X$, $\mu_{IR}(x_i, x_i) = 1$ and $\nu_{IR}(x_i, x_i) = 0$.

An intuitionistic fuzzy relation $IR \in \mathrm{IFR}(X)$ is called symmetric if for all $(x_i, x_j) \in X \times X$, $\mu_{IR}(x_i, x_j) = \mu_{IR}(x_j, x_i)$ and $\nu_{IR}(x_i, x_j) = \nu_{IR}(x_j, x_i)$.

An intuitionistic fuzzy relation $IR \in \mathrm{IFR}(X)$ is transitive if $IR \ge IR \underset{\Lambda,\mathrm{P}}{\overset{\mathrm{A,B}}{\circ}} IR$.

A transitive closure of some intuitionistic fuzzy relation $IR \in \mathrm{IFR}(X)$, is the minimum intuitionistic fuzzy relation $\tilde{IR} \in \mathrm{IFR}(X)$ which contains IR and is transitive. So, the condition $IR \le \tilde{IR}$ is met.

An intuitionistic fuzzy relation IT in X is called an intuitionistic fuzzy tolerance if it is reflexive and symmetric.

An intuitionistic fuzzy relation IS in X is called an intuitionistic fuzzy similarity relation if it is reflexive, symmetric and transitive.

An n-step procedure by using the composition of intuitionistic fuzzy relations beginning with an intuitionistic fuzzy tolerance can be used for the construction of the transitive closure of an intuitionistic fuzzy tolerance IT, and the transitive closure is an intuitionistic fuzzy similarity relation IS. This procedure is the basis of the clustering procedure which was proposed by Hung, Lee and Fuh [52].

Definition 3.6. *An α, β -level set of an intuitionistic fuzzy relation IR in X is defined as follows:*

$$IR_{\alpha,\beta} = \{(x_i, x_j) \mid \mu_{IR}(x_i, x_j) \geq \alpha, \nu_{IR}(x_i, x_j) \leq \beta\}, \tag{3.19}$$

where the condition (3.10) is met for any values α and β, $\alpha, \beta \in [0,1]$.

So, if $0 \leq \alpha_1 \leq \alpha_2 \leq 1$ and $0 \leq \beta_2 \leq \beta_1 \leq 1$, with $0 \leq \alpha_1 + \beta_1 \leq 1$ and $0 \leq \alpha_2 + \beta_2 \leq 1$, then $IR_{\alpha_2,\beta_2} \subseteq IR_{\alpha_1,\beta_1}$.

The concept of the α -level fuzzy relation in the sense of definition 1.17 can be extended to the intuitionistic fuzzy relations. So, the (α, β) -level intuitionistic fuzzy relation $IR_{(\alpha,\beta)}$ in X can be defined in the following way [130].

Definition 3.7. *Let X be a finite universe and IR be an intuitionistic fuzzy relation on X with $\mu_{IR}(x_i, x_j)$ being its membership function and $\nu_{IR}(x_i, x_j)$ being its non-membership function. The (α, β) -level intuitionistic fuzzy relation $IR_{(\alpha,\beta)}$ of the intuitionistic fuzzy relation IR on X is defined as*

$$IR_{(\alpha,\beta)} = \left\{\left\langle (x_i, x_j) \in IR_{\alpha,\beta}, \mu_{IR_{(\alpha,\beta)}}(x_i, x_j) = \mu_{IR}(x_i, x_j), \nu_{IR_{(\alpha,\beta)}}(x_i, x_j) = \nu_{IR}(x_i, x_j) \right\rangle\right\},$$

where $\alpha, \beta \in [0,1]$ satisfies the condition (3.10) and $IR_{\alpha,\beta}$ is the α, β -level of an intuitionistic fuzzy relation IR which satisfies the condition (3.19).

The concept of an (α, β) -level intuitionistic fuzzy relation is very important for our further considerations.

3.1.3 A Method of the Decomposition of Intuitionistic Fuzzy Tolerances

Let us remind a method of decomposition of intuitionistic fuzzy tolerances given in [130]. Let IT be an intuitionistic fuzzy tolerance in X. Let $IT_{(\alpha,\beta)}$ be (α, β) -level intuitionistic fuzzy relation and let the condition (3.10) be met for values α and $\beta, \alpha \in (0,1]$, $\beta \in [0,1)$. Let $IT_{\alpha,\beta}$ be a α, β -level of an intuitionistic fuzzy tolerance IT in X and $IT_{\alpha,\beta}$ be the support of $IT_{(\alpha,\beta)}$. The membership function $\mu_{IT_{(\alpha,\beta)}}(x_i, x_j)$ can be defined as

$$\mu_{IT_{(\alpha,\beta)}}(x_i, x_j) = \begin{cases} \mu_{IT}(x_i, x_j), & if\ \mu_{IT}(x_i, x_j) \geq \alpha \\ 0, & otherwise \end{cases}, \tag{3.20}$$

and the non-membership function $\nu_{IT_{(\alpha,\beta)}}(x_i, x_j)$ can be defined as

$$\nu_{IT_{(\alpha,\beta)}}(x_i, x_j) = \begin{cases} \nu_{IT}(x_i, x_j), \text{ if } \nu_{IT}(x_i, x_j) \le \beta \\ 0, \quad \text{otherwise} \end{cases} . \tag{3.21}$$

Obviously, the condition $IT_{(\alpha,\beta)} \preceq IT$ is met for any intuitionistic fuzzy tolerance IT and an (α, β)-level intuitionistic fuzzy relation $IT_{(\alpha,\beta)}$ for any $\alpha \in (0,1]$, $\beta \in [0,1)$, $0 \le \alpha + \beta \le 1$. Thus, we have the proposition: if $\alpha_{\ell(\alpha)} \le \alpha_{\ell+1(\alpha)}$ and $\beta_{\ell+1(\beta)} \le \beta_{\ell(\beta)}$ with $0 \le \alpha_{\ell(\alpha)} + \beta_{\ell(\beta)} \le 1$, $0 \le \alpha_{\ell+1(\alpha)} + \beta_{\ell+1(\beta)} \le 1$, then the condition $IT_{(\alpha_{\ell+1(\alpha)}, \beta_{\ell+1(\beta)})} \preceq IT_{(\alpha_{\ell(\alpha)}, \beta_{\ell(\beta)})}$ is met. So, the ordered sequences $0 < \alpha_0 \le \ldots \le \alpha_{\ell(\alpha)} \le \ldots \le \alpha_{Z(\alpha)} \le 1$ and $0 \le \beta_{Z(\beta)} \le \ldots \le \beta_{\ell(\beta)} \le \ldots \le \beta_0 < 1$ must be constructed for the decomposition of an intuitionistic fuzzy tolerance IT.

A method for the construction of these sequences was developed in [130]. The value α_0 is calculated as

$$\alpha_0 = \min_{x_i, x_j} \mu_{IT}(x_i, x_j), \ \forall x_i, x_j \in X , \tag{3.22}$$

and the value $\alpha_{Z(\alpha)}$ is calculated as

$$\alpha_{Z(\alpha)} = \max_{x_i, x_j} \mu_{IT}(x_i, x_j), \ \forall x_i, x_j \in X . \tag{3.23}$$

The value $\alpha_{\ell(\alpha)}$ from the sequence $0 \le \alpha_0 \le \ldots \le \alpha_{\ell(\alpha)} \le \ldots \le \alpha_{Z(\alpha)} \le 1$ can be calculated by the formula

$$\alpha_{\ell(\alpha)} = \min_{x_i, x_j} \mu_{IT}(x_i, x_j), \ \mu_{IT}(x_i, x_j) \in (\alpha_q, \alpha_{Z(\alpha)}], \ \forall x_i, x_j \in X , \tag{3.24}$$

where α_q is some current value of α.

On the other hand, the value β_0 can be calculated as follows

$$\beta_0 = \max_{x_i, x_j} \nu_{IT}(x_i, x_j), \ \forall x_i, x_j \in X . \tag{3.25}$$

The formula

$$\beta_{Z(\beta)} = \min_{x_i, x_j} \nu_{IT}(x_i, x_j), \ \forall x_i, x_j \in X , \tag{3.26}$$

can be used for the calculation of the value $\beta_{Z(\beta)}$.

The value $\beta_{\ell(\beta)}$ from the sequence $0 \le \beta_{Z(\beta)} \le \ldots \le \beta_{\ell(\beta)} \le \ldots \le \beta_0 \le 1$ can be calculated as

$$\beta_{\ell(\beta)} = \max_{x_i, x_j} \nu_{IT}(x_i, x_j),\ \nu_{IT}(x_i, x_j) \in [\beta_{Z(\beta)}, \beta_q),\ \forall x_i, x_j \in X, \qquad (3.27)$$

where β_q is a current value of β.

Thus, a sequence of (α, β) -level intuitionistic fuzzy relations $IT_{(\alpha_{\ell(\alpha)}, \beta_{\ell(\beta)})}$ can be constructed for some fixed calculated value $\alpha_{\ell(\alpha)}$ and all calculated values $\beta_{\ell(\beta)}$ from the sequence $0 \le \beta_{Z(\beta)} \le \ldots \le \beta_{\ell(\beta)} \le \ldots \le \beta_0 \le 1$ which satisfy the condition $0 \le \alpha_{\ell(\alpha)} + \beta_{\ell(\beta)} \le 1$.

The proposed method for constructing the sequences $0 < \alpha_0 \le \ldots \le \alpha_{\ell(\alpha)} \le \ldots \le \alpha_{Z(\alpha)} \le 1$ and $0 \le \beta_{Z(\beta)} \le \ldots \le \beta_{\ell(\beta)} \le \ldots \le \beta_0 < 1$ is similar to the method of decomposition of the fuzzy relations based on the concept of an α -level fuzzy relation (1.56) – (1.58). This is done by using a two-step procedure:

1. Construct the sequence $0 \le \alpha_0 \le \ldots \le \alpha_{\ell(\alpha)} \le \ldots \le \alpha_{Z(\alpha)} \le 1$ of α thresholds values as follows:

    1.1 The value α_0 from the sequence $0 \le \alpha_0 \le \ldots \le \alpha_{\ell(\alpha)} \le \ldots \le \alpha_{Z(\alpha)} \le 1$ is calculated using the formula (3.22);

    1.2 The value $\alpha_{Z(\alpha)}$ is calculated using the formula (3.23);

    1.3 Let $\alpha_q := \alpha_0$;

    1.4 The following condition is verified:
    if $\alpha_q < \alpha_{Z(\alpha)}$,
    then the value $\alpha_{\ell(\alpha)}$ is calculated using the formula (3.24),
    else go to step 1.6;

    1.5 Let $\alpha_q := \alpha_{\ell(\alpha)}$ and go to step 1.4;

    1.6 The result constitutes the sequence of all calculated values $\{\alpha_q \mid q = 0, \ldots, Z(\alpha)\}$;

2. Construct the sequence $0 \le \beta_{Z(\beta)} \le \ldots \le \beta_{\ell(\beta)} \le \ldots \le \beta_0 \le 1$ of β threshold values as follows:

2.1 The value β_0 from the sequence $0 \le \beta_{Z(\beta)} \le \ldots \le \beta_{\ell(\beta)} \le \ldots \le \beta_0 \le 1$ is calculated using the formula (3.25);

2.2 The value $\beta_{Z(\beta)}$ is calculated using the formula (3.26);

2.3 Let $\beta_q := \beta_0$;

2.4 The following condition is verified:
if $\beta_q > \beta_{Z(\beta)}$
then the value $\beta_{\ell(\beta)}$ is calculated using the formula (3.27)
else go to step 2.6;

2.5 Let $\beta_q := \beta_{\ell(\beta)}$ and go to step 2.4;

2.6 The result contitutes the sequence of all calculated values $\{\beta_q \mid q = Z(\beta), \ldots, 0\}$;

The algorithm for the construction of (α, β)-level intuitionistic fuzzy relations can be described as the following five-step procedure:

1. Let $\ell(\alpha) := 0$;
2. Let $\ell(\beta) := 0$;
3. The following condition is checked:
   if the condition $\alpha_{\ell(\alpha)} + \beta_{\ell(\beta)} \le 1$ is met
   then the (α, β)-level intuitionistic fuzzy relation $IT_{(\alpha,\beta)}$ is constructed for the corresponding values $\alpha_{\ell(\alpha)}$ and $\beta_{\ell(\beta)}$ using the formulae (3.20) and (3.21)
   else the following condition is checked:
   if the condition $\beta_{\ell(\beta)} > \beta_{Z(\beta)}$ is met
   then $\ell(\beta) := \ell(\beta) + 1$ and go to step 3;
   else go to step 4;
4. The following condition is checked:
   if the condition $\alpha_{\ell(\alpha)} < \alpha_{Z(\alpha)}$ is met
   then $\ell(\alpha) := \ell(\alpha) + 1$ and go to step 2;
   else go to step 5;
5. The result constitutes the set of (α, β)-level intuitionistic fuzzy relations for each value $\alpha_{\ell(\alpha)}$ from the sequence $0 \le \alpha_0 \le \ldots \le \alpha_{\ell(\alpha)} \le \ldots \le \alpha_{Z(\alpha)} \le 1$.

So, a tree of (α, β)-level intuitionistic fuzzy relations will be constructed for some intuitionistic fuzzy tolerance IT .

The proposed method for the construction of (α, β) -level intuitionistic fuzzy relations may be explained on a simple example. Let us consider the execution of the proposed method for the matrix of the intuitionistic fuzzy tolerance IT in $X = \{x_1, x_2, x_3\}$. This matrix of the intuitionistic fuzzy tolerance originally appeared as Table 3.1 in [52].

Table 3.1 Matrix of the intuitionistic fuzzy tolerance

IT	x_1	x_2	x_3
x_1	(1.00, 0.00)		
x_2	(0.80, 0.20)	(1.00, 0.00)	
x_3	(0.70, 0.25)	(0.20, 0.75)	(1.00, 0.00)

The first algorithm has been applied to this matrix. So, the values α can be ordered as follows: $0.20 \le 0.70 \le 0.80 \le 1.00$. On the other hand, an ordered sequence $0.00 \le 0.20 \le 0.25 \le 0.75$ is the sequence of values β.

So, the set of (α, β) -level intuitionistic fuzzy relations has been obtained for each value $\alpha_{\ell(\alpha)}$ from the second algorithm. For example, the values $\beta_{Z(\beta)=3} = 0.00$ and $\beta_{\ell(\beta)=2} = 0.20$ satisfy the condition $0 \le \alpha_{\ell(\alpha)} + \beta_{\ell(\beta)} \le 1$ for the corresponding value $\alpha_{\ell(\alpha)=2} = 0.80$. The obtained tree of (α, β) -level intuitionistic fuzzy relations is presented in Figure 3.1.

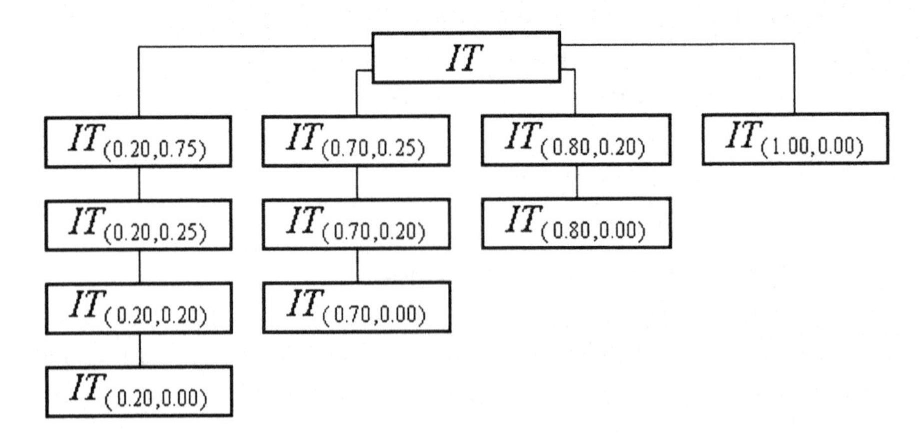

Fig. 3.1 The tree of (α,β)-level intuitionistic fuzzy relations

Obviously, the condition $IT_{(1.00,0.00)} \preceq IT_{(0.80,0.20)} \preceq IT_{(0.70,0.25)} \preceq IT_{(0.20,0.75)}$ is met. Note that the (α, β) -level intuitionistic fuzzy tolerance $IT_{(0.20,0.75)}$ is equal to the initial intuitionistic fuzzy tolerance IT .

3.1.4 Basic Concepts of an Intuitionistic Fuzzy Generalization of the Clustering Approach

An intuitionistic fuzzy generalization of the D-PAFC-algorithm was proposed in [140]. The corresponding D-PAIFC-algorithm is based on the intuitionistic fuzzy relations and aims at the detection of an unknown least number of compact and well-separated intuitionistic fuzzy clusters. Let us consider basic concepts of the D-PAIFC-algorithm.

Let $X = \{x_1, \ldots, x_n\}$ be the initial set of elements and IT be some binary intuitionistic fuzzy tolerance on $X = \{x_1, \ldots, x_n\}$ with $\mu_{IT}(x_i, x_j) \in [0,1]$ being its membership function and $\nu_{IT}(x_i, x_j) \in [0,1]$ being its non-membership function. Let α and β be the α, β -level values of IT , $\alpha \in (0,1]$, $\beta \in [0,1)$, $0 \le \alpha + \beta \le 1$. The columns or rows of the intuitionistic fuzzy tolerance matrix are intuitionistic fuzzy sets $\{IA^1, \ldots, IA^n\}$. Let $\{IA^1, \ldots, IA^n\}$ on X which are generated by an intuitionistic fuzzy tolerance IT .

Definition 3.8. *Let $\{IA^1, \ldots, IA^n\}$ be intuitionistic fuzzy sets on X which are generated by an intuitionistic fuzzy tolerance IT . The (α, β) -level intuitionistic fuzzy set $IA^l_{(\alpha,\beta)} = \{(x_i, \mu_{IA^l}(x_i), \nu_{IA^l}(x_i)) \mid \mu_{IA^l}(x_i) \ge \alpha, \nu_{IA^l}(x_i) \le \beta, x_i \in X\}$ is an intuitionistic fuzzy (α, β) -cluster or, simply, an intuitionistic fuzzy cluster. So, $IA^l_{(\alpha,\beta)} \subseteq IA^l$, $\alpha \in (0,1]$, $\beta \in [0,1)$, $IA^l \in \{IA^1, \ldots, IA^n\}$ and μ_{li} is the membership degree of the element $x_i \in X$ for some intuitionistic fuzzy cluster $IA^l_{(\alpha,\beta)}$, $\alpha \in (0,1]$, $\beta \in [0,1)$, $l \in \{1, \ldots, n\}$. On the other hand, ν_{li} is the non-membership degree of the element $x_i \in X$ for the intuitionistic fuzzy cluster $IA^l_{(\alpha,\beta)}$. The value of α is a tolerance threshold of elements of the intuitionistic fuzzy clusters and the value of β is the difference threshold of elements of the intuitionistic fuzzy clusters.*

The membership degree of the element $x_i \in X$ for some intuitionistic fuzzy cluster $IA^l_{(\alpha,\beta)}$, $\alpha \in (0,1]$, $\beta \in [0,1)$, $0 \le \alpha + \beta \le 1$, $l \in \{1, \ldots, n\}$, is defined as

$$\mu_{li} = \begin{cases} \mu_{IA^l}(x_i), & x_i \in IA^l_{\alpha,\beta} \\ 0, & otherwise \end{cases}, \tag{3.28}$$

where an α, β-level $IA^l_{\alpha,\beta}$ of an intuitionistic fuzzy set IA^l is the support of the intuitionistic fuzzy cluster $IA^l_{(\alpha,\beta)}$. So, the condition $IA^l_{\alpha,\beta} = Supp(IA^l_{(\alpha,\beta)})$ is met for each intuitionistic fuzzy cluster $IA^l_{(\alpha,\beta)}$. The membership degree μ_{li} can be interpreted as a degree of typicality of an element to an intuitionistic fuzzy cluster.

On the other hand, the non-membership degree of the element $x_i \in X$ for an intuitionistic fuzzy cluster $IA^l_{(\alpha,\beta)}$, $\alpha \in (0,1]$, $\beta \in [0,1)$, $0 \le \alpha + \beta \le 1$, $l \in \{1,\ldots,n\}$, can be defined as

$$\nu_{li} = \begin{cases} \nu_{A^l}(x_i), & x_i \in IA^l_{\alpha,\beta} \\ 0, & otherwise \end{cases}. \tag{3.29}$$

So, the non-membership degree ν_{li} can be interpreted as a degree of non-typicality of an element to an intuitionistic fuzzy cluster.

Thus, if the columns or rows of an intuitionistic fuzzy tolerance IT matrix are intuitionistic fuzzy sets $\{IA^1,\ldots,IA^n\}$ on X, then the intuitionistic fuzzy clusters $\{IA^1_{(\alpha,\beta)},\ldots,IA^n_{(\alpha,\beta)}\}$ are intuitionistic fuzzy subsets of fuzzy sets $\{IA^1,\ldots,IA^n\}$ for some pair of values $\alpha \in (0,1]$ and $\beta \in [0,1)$, $0 \le \alpha + \beta \le 1$. So, the condition $0 \le \mu_{li} + \nu_{li} \le 1$ is met for some intuitionistic fuzzy cluster $IA^l_{(\alpha,\beta)}$.

Definition 3.9. *If the conditions* $\mu_{li} = 0$ *and* $\nu_{li} = 0$ *are met for some element* $x_i \in X$ *and for some intuitionistic fuzzy cluster* $IA^l_{(\alpha,\beta)}$, *then this element* $x_i \in X$ *is called the residual element of the intuitionistic fuzzy cluster* $IA^l_{(\alpha,\beta)}$.

The value equal zero for a fuzzy set membership function is equivalent to the non-belongingness of an element to a fuzzy set. That is why values of the tolerance threshold α are considered to be from the interval $(0,1]$. So, the value of the membership function of each element of the intuitionistic fuzzy cluster is the degree of similarity of the object to some typical object of the fuzzy cluster. On the other hand, the value equal one for an intuitionistic fuzzy set non-membership function is equivalent to the non-belongingness of an element to theintuitionistic fuzzy set. That is why values of the difference threshold β are considered to be from the interval $[0,1)$.

Definition 3.10. *Let IT is an intuitionistic fuzzy tolerance on X, where X is the set of elements, and $\{IA^1_{(\alpha,\beta)},\ldots,IA^n_{(\alpha,\beta)}\}$ is a family of intuitionistic fuzzy clusters for some $\alpha\in(0,1]$ and $\beta\in[0,1)$. The point $\tau^l_e\in IA^l_{\alpha,\beta}$, for which*

$$\tau^l_e = \arg\max_{x_i} \mu_{li}\,,\ \forall x_i \in IA^l_{\alpha,\beta} \tag{3.30}$$

is called a typical point of the intuitionistic fuzzy cluster $IA^l_{(\alpha,\beta)}$.

The membership degree of a typical point of an intuitionistic fuzzy cluster is equal one because an intuitionistic fuzzy tolerance IT is a reflexive intuitionistic fuzzy relation. So, the non-membership degree of a typical point of an intuitionistic fuzzy cluster should be equal zero. Moreover, a typical point of an intuitionistic fuzzy cluster does not depend on the value of tolerance threshold and an intuitionistic fuzzy cluster can have several typical points. That is why the symbol e is the index of the typical point.

Definition 3.11. *Let $IR^{\alpha,\beta}_z(X)=\{IA^l_{(\alpha,\beta)} \mid l=\overline{1,c},\ c\le n\}$ be a family of intuitionistic fuzzy clusters for some value of a tolerance threshold $\alpha\in(0,1]$ and some value of a difference threshold $\beta\in[0,1)$, $0\le\alpha+\beta\le 1$. These intuitionistic fuzzy clusters are generated by some intuitionistic fuzzy tolerance IT on the initial set of elements $X=\{x_1,\ldots,x_n\}$.*

If the condition

$$\sum_{l=1}^{c}\mu_{li} > 0\,,\ \forall x_i \in X \tag{3.31}$$

and the condition

$$\sum_{l=1}^{c}\nu_{li} \ge 0\,,\ \forall x_i \in X \tag{3.32}$$

are met for all $IA^l_{(\alpha,\beta)}$, $l=\overline{1,c}$, $c\le n$, then this family is an allotment of elements of the set $X=\{x_1,\ldots,x_n\}$ among intuitionistic fuzzy clusters $\{IA^l_{(\alpha,\beta)}, l=\overline{1,c}, 2\le c\le n\}$ for some value of the tolerance threshold $\alpha\in(0,1]$ and some value of the difference threshold $\beta\in[0,1)$.

It should be noted that several allotments $IR^{\alpha,\beta}_z(X)$ can exist for the same pair of thresholds α and β. That is why symbol z is the index of an allotment.

The condition (3.31) requires that each object $x_i, i=\overline{1,n}$ must be assigned to at least one intuitionistic fuzzy cluster $IA^l_{(\alpha,\beta)}$, $l=\overline{1,c}$, $c \leq n$, with the membership degree higher than zero and this condition is similar to the definition of the possibilistic partition [66]. The condition $2 \leq c \leq n$ requires that the number of intuitionistic fuzzy clusters in $IR^{\alpha,\beta}_z(X)$ be more than two. Otherwise, the unique intuitionistic fuzzy cluster will contain all objects, possibly with different positive membership and non-membership degrees.

Definition 3.12. *An allotment* $IR^{\alpha,\beta}_I(X)=\{IA^l_{(\alpha,\beta)} \mid l=\overline{1,n}, \alpha \in (0,1], \beta \in [0,1)\}$ *of the set of objects among* n *intuitionistic fuzzy clusters for some pair of thresholds* α *and* β, $0 \leq \alpha+\beta \leq 1$, *is an initial allotment of the set* $X=\{x_1,\ldots,x_n\}$.

In other words, if the initial data are represented by a matrix of some intuitionistic fuzzy tolerance relation IT, then the rows or columns of the matrix are intuitionistic fuzzy sets IA^l, $l=\overline{1,n}$, and (α,β)-level fuzzy sets $IA^l_{(\alpha,\beta)}$, $l=\overline{1,n}$, $\alpha \in (0,1]$, $\beta \in [0,1)$, are intuitionistic fuzzy clusters. These intuitionistic fuzzy clusters constitute an initial allotment for some pair of thresholds α and β, and they can be considered as clustering components.

Definition 3.13. *If the condition*

$$\bigcup_{l=1}^{c} IA^l_{\alpha,\beta} = X, \tag{3.33}$$

and the condition

$$card(IA^l_{\alpha,\beta} \cap IA^m_{\alpha,\beta})=0, \ \forall IA^l_{(\alpha,\beta)},\ IA^m_{(\alpha,\beta)},\ l \neq m,\ \alpha,\beta \in (0,1] \tag{3.34}$$

are met for all intuitionistic fuzzy clusters $IA^l_{(\alpha,\beta)}$, $l=\overline{1,c}$ *of some allotment* $IR^{\alpha,\beta}_z(X)=\{IA^l_{(\alpha,\beta)} \mid l=\overline{1,c}, c \leq n, \alpha \in (0,1], \beta \in [0,1)\}$, *then this allotment is the allotment among fully separated intuitionistic fuzzy clusters.*

Thus, the problem of cluster analysis can here be defined in general as the problem of discovering an unknown least number of compact and well-separated intuitionistic fuzzy clusters. For this purpose, the concept of a principal allotment among intuitionistic fuzzy clusters is introduced as follows.

Definition 3.14. *An allotment* $IR_P^{\alpha,\beta}(X)=\{IA_{(\alpha,\beta)}^l \mid l=\overline{1,c}, c\le n, \alpha,\beta\in[0,1]\}$ *of the set of objects among the minimal number* $2\le c<n$ *of fully separated intuitionistic fuzzy clusters, for some value of the tolerance threshold* $\alpha\in(0,1]$ *and some value of the difference threshold* $\beta\in[0,1)$, *is a principal allotment of the set* $X=\{x_1,\ldots,x_n\}$.

Several principal allotments can exist for some pair of the thresholds α and β. Thus, the problem consists in the selection of the unique principal allotment $IR^*(X)$ from the set B of principal allotments, $B=\{IR_{P_z}^{\alpha,\beta}(X)\}$, which is the class of possible solutions of the specific classification problem. The symbol z is the index of the principal allotment.

The selection of the unique principal allotment $IR^*(X)$ from the set $B=\{IR_{P_z}^{\alpha,\beta}(X)\}$ of principal allotments is made on the basis of evaluation of allotments. One can use the criterion

$$F(IR_{P_z}^{\alpha,\beta}(X),\alpha,\beta)=\left(\sum_{l=1}^{c}\frac{1}{n_l}\sum_{i=1}^{n_l}\mu_{li}-\alpha\cdot c\right)-\left(\sum_{l=1}^{c}\frac{1}{n_l}\sum_{i=1}^{n_l}\nu_{li}-\beta\cdot c\right), \quad (3.35)$$

where c is the number of intuitionistic fuzzy clusters in the allotment $IR_{P_z}^{\alpha,\beta}(X)$ and $n_l=card(IA_{\alpha,\beta}^l)$, $IA_{(\alpha,\beta)}^l\in IR_{P_z}^{\alpha,\beta}(X)$ is the number of elements in the support of the intuitionistic fuzzy cluster $IA_{(\alpha,\beta)}^l$. The criterion (3.35) is an intuitionistic fuzzy extension of the criterion (2.8).

The maximum of the criterion (3.35) corresponds to the best allotment of objects among c intuitionistic fuzzy clusters. Thus, the classification problem can be characterized formally as the determination of a solution $IR^*(X)$ satisfying

$$IR^*(X)=\arg\max_{IR_z^{\alpha,\beta}(X)\in B} F(IR_z^{\alpha,\beta}(X),\alpha,\beta), \quad (3.36)$$

where $B=\{IR_{P_z}^{\alpha,\beta}(X)\}$ is the set of principal allotments among intuitionistic fuzzy clusters corresponding to the pair of thresholds α and β.

3.1.5 An Outline of the D-PAIFC-Algorithm

The principal allotment $IR_P^{\alpha,\beta}(X)$ is a family of intuitionistic fuzzy clusters which are elements of the initial allotment $IR_I^{\alpha,\beta}(X)$ for the pair of thresholds α

and β; the family of intuitionistic fuzzy clusters should satisfy the conditions (3.33) and (3.34). So, the construction of the principal allotments $IR_{P_z}^{\alpha,\beta}(X) = \{IA_{(\alpha,\beta)}^l \mid l = \overline{1,c}, c \le n, \alpha \in (0,1], \beta \in [0,1)\}$ for each pair of thresholds α and β is a combinatorial problem.

There exists a six-step clustering algorithm for the detection of the principal allotment among intuitionistic fuzzy clusters as shown below.

The D-PAIFC-Algorithm

1. Construct the ordered sequences of threshold values α and β, $0 < \alpha_0 \le \ldots \le \alpha_{\ell(\alpha)} \le \ldots \le \alpha_{Z(\alpha)} \le 1$ and $0 \le \beta_{Z(\beta)} \le \ldots \le \beta_{\ell(\beta)} \le \ldots \le \beta_0 < 1$; let $\ell(\alpha) := 0$ and $\ell(\beta) := 0$;
2. Check the following condition:

   if the condition $0 \le \alpha_{\ell(\alpha)} + \beta_{\ell(\beta)} \le 1$ is met

   then construct the (α,β)-level intuitionistic fuzzy relation $IT_{(\alpha,\beta)}$ in the sense of definition 3.7 and go to Step 3

   else check the following condition:

   if the condition $\ell(\beta) < Z(\beta)$ is met

   then let $\ell(\beta) := \ell(\beta) + 1$ and go to step 2;
3. Construct the initial allotment $IR_I^{\alpha,\beta}(X) = \{IA_{(\alpha,\beta)}^l \mid l = \overline{1,n}, \alpha \in (0,1], \beta \in [0,1)\}$ for the calculated values $\alpha_{\ell(\alpha)}$ and $\beta_{\ell(\beta)}$;
4. The following condition is checked:

   if for some intuitionistic fuzzy cluster $IA_{(\alpha,\beta)}^l \in IR_I^{\alpha,\beta}(X)$ the condition $card(IA_{(\alpha,\beta)}^l) = n$ is met

   then let $\ell(\beta) := \ell(\beta) + 1$ and go to step 2

   else construct the allotments which satisfy the conditions (3.33) and (3.34);
5. The following condition is checked:

   if for $\alpha_{\ell(\alpha)}$ and $\beta_{\ell(\beta)}$ the allotments $IR_z^{\alpha}(X)$ satisfying the conditions (3.33) and (3.34) are not constructed and the condition $\ell(\beta) < Z(\beta)$ is met

   then let $\ell(\beta) := \ell(\beta) + 1$ and go to step 2

   else **if** the condition $\ell(\beta) = Z(\beta)$ is met

   then $\ell(\alpha) := \ell(\alpha) + 1$ and go to step 2

   else construct the class of possible solutions of the classification problem $B = \{IR_{P_z}^{\alpha,\beta}(X)\}$ for $\alpha_{\ell(\alpha)}$ and $\beta_{\ell(\beta)}$;

6. The following condition is checked:

   if condition $card(B) > 1$ is met

   then calculate the value of the criterion (3.35) for each allotment $IR_{P_z}^{\alpha,\beta}(X) \in B$

   and the result of classification, $IR^*(X)$, is constructed as follows:

   if for some unique allotment $IR_{P_z}^{\alpha,\beta}(X) \in B$ the condition (3.36) is met

   then the allotment is a solution $IR^*(X)$ of the classification problem;

   else **if the** condition $card(B) = 1$ is met

   then the unique allotment $IR_P^{\alpha,\beta}(X) \in B$ is a solution $IR^*(X)$ of the classification problem.

The unique principal allotment $IR_P^{\alpha,\beta}(X)$ among an unknown least number of fully separate intuitionistic fuzzy clusters and the corresponding values of the tolerance threshold α and the difference threshold β, $0 \leq \alpha + \beta \leq 1$ are results of classification.

3.1.6 Experimental Results

Let us consider an application of the proposed D-PAIFC-algorithm to the classification problem. An intuitionistic fuzzy tolerance relation matrix is given in Table 3.2.

Table 3.2 Intuitionistic fuzzy tolerance relation matrix

IT	x_1	x_2	x_3	x_4	x_5	x_6	x_7	x_8	x_9	x_{10}
x_1	(1.0, 0.0)									
x_2	(0.2, 0.7)	(1.0, 0.0)								
x_3	(0.5, 0.5)	(0.3, 0.6)	(1.0, 0.0)							
x_4	(0.8, 0.1)	(0.6, 0.4)	(0.5, 0.4)	(1.0, 0.0)						
x_5	(0.6, 0.3)	(0.7, 0.2)	(0.3, 0.6)	(0.7, 0.2)	(1.0, 0.0)					
x_6	(0.2, 0.7)	(0.9, 0.1)	(0.4, 0.5)	(0.3, 0.6)	(0.2, 0.7)	(1.0, 0.0)				
x_7	(0.3, 0.7)	(0.2, 0.7)	(0.1, 0.9)	(0.5, 0.4)	(0.4, 0.5)	(0.1, 0.7)	(1.0, 0.0)			
x_8	(0.9, 0.1)	(0.8, 0.2)	(0.3, 0.6)	(0.4, 0.6)	(0.5, 0.5)	(0.3, 0.7)	(0.6, 0.3)	(1.0, 0.0)		
x_9	(0.4, 0.5)	(0.3, 0.7)	(0.7, 0.2)	(0.1, 0.8)	(0.8, 0.1)	(0.7, 0.2)	(0.1, 0.8)	(0.0, 0.9)	(1.0, 0.0)	
x_{10}	(0.3, 0.7)	(0.2, 0.7)	(0.6, 0.3)	(0.3, 0.7)	(0.9, 0.1)	(0.2, 0.7)	(0.3, 0.7)	(0.2, 0.8)	(0.1, 0.8)	(1.0, 0.0)

Let us consider the classification result which was presented by Hung, Lee and Fuh in [52] where these data have originally appeear. An intuitionistic fuzzy similarity relation matrix $IS = [\mu_{IS}(x_i, x_j), \nu_{IS}(x_i, x_j)]$, $i, j = 1, \ldots, 10$, was obtained by the n-step $\max - T_3$ & $\min - S_3$ composition procedure. A hard partition is a result of he application of their method to the intuitionistic fuzzy similarity relation IS. The partition $\{x_1, x_4, x_8\}$, $\{x_2, x_5, x_9, x_{10}\}$, $\{x_3\}$, $\{x_6\}$, $\{x_7\}$ was obtained for $\alpha = 0.55$ and $\beta = 0.35$. For comparison, the D-PAIFC-algorithm was directly applied to the intuitionistic fuzzy tolerance matrix IT. Let us consider the result of the experiment.

By executing the D-PAIFC-algorithm we obtain a class of possible solutions of the classification problem $B = \{IR_{P_1}^{\alpha,\beta}(X), IR_{P_2}^{\alpha,\beta}(X)\}$, for $\alpha = 0.8$ and $\beta = 0.2$. The first principal allotment is the family $IR_{P_1}^{\alpha,\beta}(X) = \{IA_{(\alpha,\beta)}^1, IA_{(\alpha,\beta)}^3, IA_{(\alpha,\beta)}^5, IA_{(\alpha,\beta)}^6, IA_{(\alpha,\beta)}^7\}$ and the second principal allotment is the family $IR_{P_2}^{\alpha,\beta}(X) = \{IA_{(\alpha,\beta)}^2, IA_{(\alpha,\beta)}^3, IA_{(\alpha,\beta)}^4, IA_{(\alpha,\beta)}^5, IA_{(\alpha,\beta)}^7\}$, where:

$$IA_{(\alpha,\beta)}^1 = \{(x_1, 1.0, 0.0), (x_4, 0.8, 0.1), (x_8, 0.9, 0.1)\},$$
$$IA_{(\alpha,\beta)}^2 = \{(x_2, 0.9, 0.1), (x_6, 1.0, 0.0), (x_8, 0.8, 0.2)\},$$
$$IA_{(\alpha,\beta)}^3 = \{(x_3, 1.0, 0.0)\},$$
$$IA_{(\alpha,\beta)}^4 = \{(x_1, 1.0, 0.0), (x_4, 0.8, 0.1)\},$$
$$IA_{(\alpha,\beta)}^5 = \{(x_5, 1.0, 0.0), (x_9, 0.8, 0.1), (x_{10}, 0.9, 0.1)\},$$
$$IA_{(\alpha,\beta)}^6 = \{(x_2, 0.9, 0.1), (x_6, 1.0, 0.0)\},$$
$$IA_{(\alpha,\beta)}^7 = \{(x_7, 1.0, 0.0)\}$$

are intuitionistic fuzzy clusters. So, both the principal allotments among intuitionistic fuzzy clusters are possible solutions of the classification problem and a unique principal allotment should be chosen as the solution $IR^*(X)$. The value of the criterion (3.35) is equal 1.5844 for the principal allotment $IR_{P_1}^{\alpha,\beta}(X) \in B$ and the value of the criterion (3.35) is equal 1.4834 for the principal allotment $IR_{P_2}^{\alpha,\beta}(X) \in B$. So, the first principal allotment is the solution $IR^*(X)$ of the classification problem.

The matrix of principal allotment among intuitionistic fuzzy clusters can be illustrated by a diagram. The membership values and non-membership values of five classes, obtained from the D-PAIFC-algorithm, are presented in Figure 3.2.

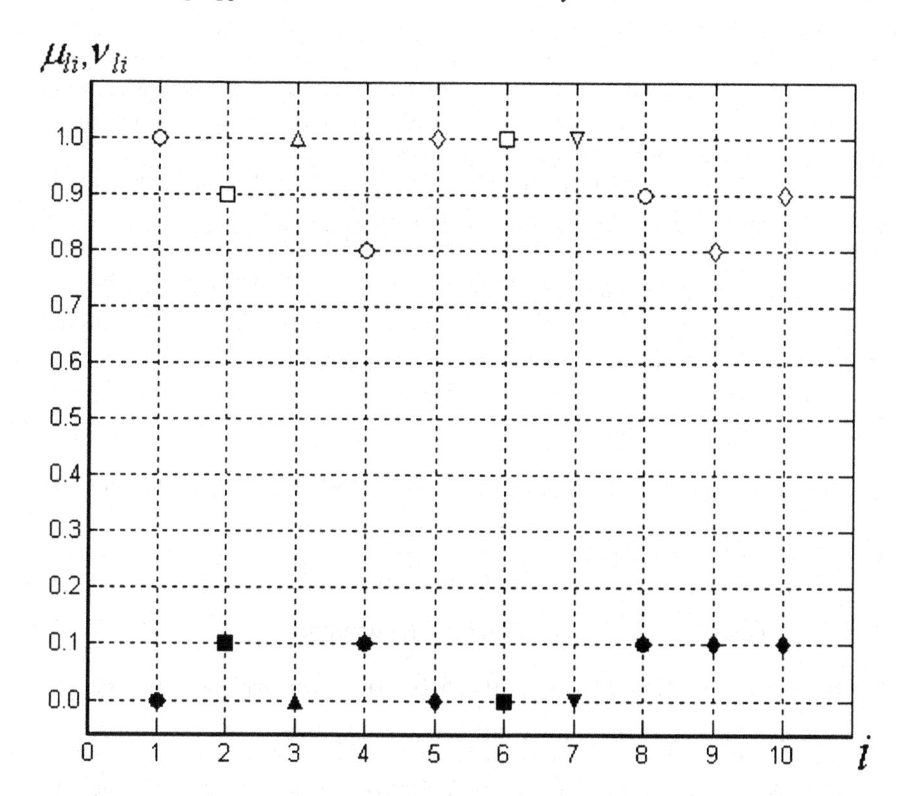

Fig. 3.2 Membership values and non-membership values of five intuitionistic fuzzy clusters

The membership values of the first class are represented in Figure 3.2 by ○, the membership values of the second class are represented by □, the membership values of the third class are represented by Δ, the membership values of the fourth class are represented by ∇, and the membership values of the fifth class are represented by ◊. The non-membership values of the first class are represented by ●, the non-membership values of the second class are represented by ■, the non-membership values of the third class are represented by ▲, the non-membership values of the fourth class are represented by ▼, and the non-membership values of the fifth class are represented by ♦. The membership values and non-membership values of residual elements of the intuitionistic fuzzy clusters are not shown in Figure 3.2.

Therefore, we obtain the following: the first class is composed of three elements, the second class is composed of two elements, the third class is composed of one element, the fourth class contains one element, and the fifth class is composed of three elements. The first object is the a typical point of the fuzzy cluster which corresponds to the first class; the sixth object is the typical point of the second fuzzy cluster; the fifth object is the typical point of the fifth fuzzy cluster. Obviously, the third object is the typical point of the fuzzy cluster which corresponds to the third class and the seventh object is the typical point of the fuzzy cluster which corresponds to the fourth class.

Let us consider an application of the proposed D-PAIFC-algorithm to the classification problem for the second illustrative example which was also considered by Hung, Lee and Fuh [52]. Seven different figures are presented in Figure 3.3 and each figure is denoted by x_i, $i \in \{1,\ldots,7\}$. These figures have all the basic pattern Δ and the pattern is denoted by x_1 in Figure 3.3. Different crossing lines have been added to the basic pattern in each case: “–”, “ | ”, “|”, “|”. Their subjective relations are stated in [52] as follows:

A. $(\mu_{IT}(x_i,x_j), \nu_{IT}(x_i,x_j)) = (0.75, 0.20)$ when patterns x_i and x_j, $i, j \in \{1,\ldots,7\}$, have only one different crossing line;

B. $(\mu_{IT}(x_i,x_j), \nu_{IT}(x_i,x_j)) = (0.50, 0.50)$ when patterns x_i and x_j, $i, j \in \{1,\ldots,7\}$, have two different crossing line;

C. $(\mu_{IT}(x_i,x_j), \nu_{IT}(x_i,x_j)) = (0.25, 0.70)$ when patterns x_i and x_j, $i, j \in \{1,\ldots,7\}$, have three different crossing line;

D. $(\mu_{IT}(x_i,x_j), \nu_{IT}(x_i,x_j)) = (0.00, 0.90)$ when patterns x_i and x_j, $i, j \in \{1,\ldots,7\}$, have four different crossing line.

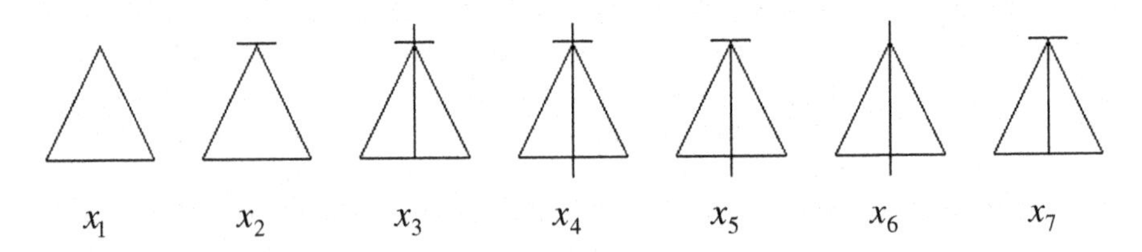

Fig. 3.3 Seven different figures

So, $(\mu_{IT}(x_i,x_i), \nu_{IT}(x_i,x_i)) = (1.00, 0.00)$ for the same pattern x_i, $i \in \{1,\ldots,7\}$. These relations assigned to the pairs of figures collected are presented in the tabular form in Table 3.3.

Table 3.3 Matrix of (subjective) intuitionistic fuzzy relations

IT	x_1	x_2	x_3	x_4	x_5	x_6	x_7
x_1	(1.00, 0.00)						
x_2	(0.75, 0.20)	(1.00, 0.00)					
x_3	(0.25, 0.70)	(0.50, 0.50)	(1.00, 0.00)				
x_4	(0.00, 0.90)	(0.25, 0.70)	(0.75, 0.20)	(1.00, 0.00)			
x_5	(0.25, 0.70)	(0.50, 0.50)	(0.50, 0.50)	(0.75, 0.20)	(1.00, 0.00)		
x_6	(0.25, 0.70)	(0.00, 0.90)	(0.50, 0.50)	(0.75, 0.20)	(0.50, 0.50)	(1.00, 0.00)	
x_7	(0.50, 0.50)	(0.75, 0.20)	(0.75, 0.20)	(0.50, 0.50)	(0.75, 0.20)	(0.25, 0.70)	(1.00, 0.00)

Hung, Lee and Fuh [52] applied their algorithm to the data using the n-step $\max\text{–}T_1$ & $\min\text{–}S_1$ composition procedure and the n-step $\max\text{–}T_3$ & $\min\text{–}S_3$ composition procedure. If the n-step $\max\text{–}T_1$ & $\min\text{–}S_1$ composition procedure is chosen, then the partition $\{x_1, x_2, x_3, x_4, x_5, x_6, x_7\}$ was obtained for $0 < \alpha \leq 0.75$ and $0 < \beta \leq 0.20$. The partition $\{x_1\}$, $\{x_2\}$, $\{x_3\}$, $\{x_4\}$, $\{x_5\}$, $\{x_6\}$, $\{x_7\}$ was obtained for $0.75 < \alpha \leq 1.00$ and $0 < \beta \leq 0.20$, $0.20 < \beta \leq 1.00$ under constraints (3.10). On the other hand, some variants of the result obtained with the n-step $\max\text{–}T_3$ & $\min\text{–}S_3$ composition procedure are presented in Table 3.4.

Table 3.4 Hard partitions obtained by Hung, Lee and Fuh's [52] algorithm

Values of the thresholds α and β	Hard partition
$0.00 < \alpha \leq 0.25$, $0.00 \leq \beta < 0.20$	$\{x_1\}$, $\{x_2\}$, $\{x_3\}$, $\{x_4\}$, $\{x_5\}$, $\{x_6\}$, $\{x_7\}$
$0.00 < \alpha \leq 0.25$, $0.20 \leq \beta < 0.40$	$\{x_1, x_2\}$, $\{x_3, x_4, x_7\}$, $\{x_5\}$, $\{x_6\}$ or $\{x_1, x_2\}$, $\{x_4, x_5, x_7\}$, $\{x_3\}$, $\{x_6\}$ or...
$0.00 < \alpha \leq 0.25$, $0.40 \leq \beta < 0.60$	$\{x_1, x_2, x_7\}$, $\{x_3, x_4, x_5, x_6\}$ or $\{x_3, x_4, x_5, x_6, x_7\}$, $\{x_1, x_2\}$ or...
...	...

So, a set of hard partitions can be obtained for difference values of the thresholds α and β under constraints (3.10).

Hung, Lee and Fuh [52] note that the clustering results based on the $\max\text{–}T_3$ & $\min\text{–}S_3$ compositions is "softer" than the clustering results based on the $\max - T_1$ & $\min - S_1$ compositions. Moreover, they note that the $\max - T_3$ & $\min - S_3$ compositions seem to have clustering results more acceptable than those of the $\max - T_1$ & $\min - S_1$ compositions. However, no method for the selection of a unique hard partition from the set of hard partitions was proposed in [52].

In fact, the matrix of subjective relations is an intuitionistic fuzzy tolerance matrix. That is why the D-PAIFC-algorithm can be applied to this matrix directly. After the application of the D-PAIFC-algorithm to the matrix of initial data, the principal allotment $IR_P^{\alpha,\beta}(X)$ among two intuitionistic fuzzy clusters, which corresponds to the classification result, is obtained for $\alpha = 0.25$ and $\beta = 0.5$.

The membership and non-membership degrees of two classes of the principal allotment $IR_P^{\alpha,\beta}(X)$, obtained from the D-PAIFC-algorithm, are presented in Figure 3.4. The membership values of the first class are represented in Figure 3.4 by ○ and the membership values of the second class are represented by □. The non-membership values of the first class are represented in Figure 3.4 by ■ and the non-membership values of the second class are represented by ●.

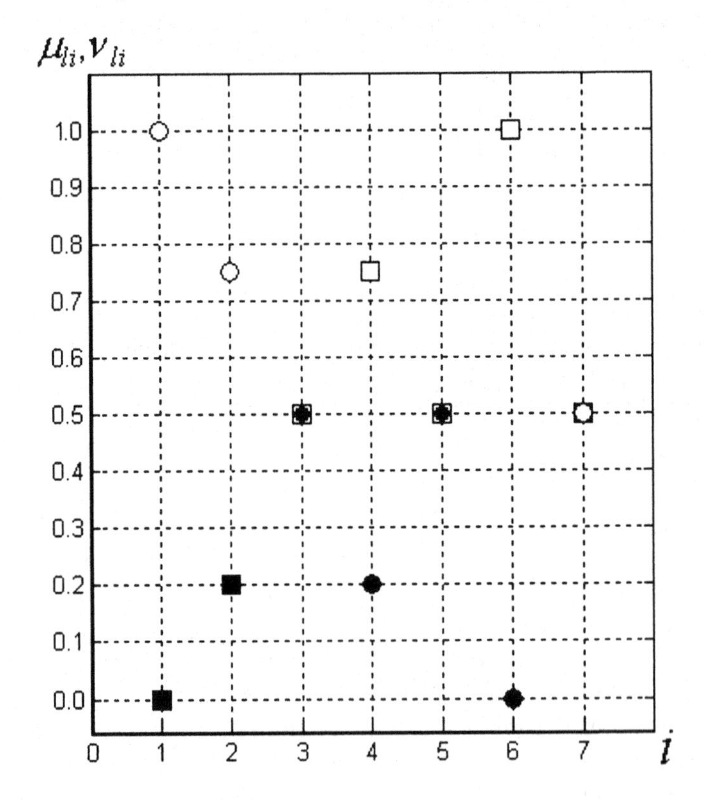

Fig. 3.4 The membership values and non-membership valuesof two intuitionistic fuzzy clusters

The first class of the principal allotment includes three elements and the second class is formed by four elements. The first object is the typical point of the first fuzzy cluster and the sixth object is the typical point of the second fuzzy cluster.

The results of application of the proposed D-PAIFC-algorithm to Hung, Lee and Fuh's [52] data sets show that the D-PAIFC-algorithm is a precise and numerically effective and efficient procedure for solving the classification problems. The membership values and non-membership values correspond to those elements of the intuitionistic fuzzy clusters which are elements of a principal allotment. On the other hand, a hard partition is the result of classification obtained from Hung, Lee and Fuh's clustering algorithm, as well as from other clustering procedures based on the intuitionistic fuzzy relations. So, a comparative study with Hung, Lee and Fuh's [52] clustering algorithm shows that the D-PAIFC-algorithm is more reasonable than other relational clustering algorithms. The concepts of an intuitionistic fuzzy cluster and allotment have an epistemological motivation. That is why the results of application of the proposed clustering method based on the allotment concept can be very well interpreted. Moreover, the clustering method based on the allotment concept depends on the set B of principal allotments only. That is why the clustering results are stable.

In conclusion it should be said that some parameters can be introduced into a clustering procedure [140]. In particular, the detection of a fixed number of fuzzy clusters c, can be considered as the purpose of classification. The intuitionistic fuzzy clusters can have a non-empty intersection area and such intuitionistic fuzzy clusters are called particularly separate fuzzy clusters in the case. So, the conditions (3.33) and (3.34) can be generalized to the case of particularly separate intuitionistic fuzzy clusters. The number c of intuitionistic fuzzy clusters in the allotment $IR^{*}(X)$ sought can be considered as a parameter of a modification of the clustering procedure. Moreover, some other parameters are considered in [140]. On the other hand, the proposed approach can be considered as a basis for the elaboration of hierarchical clustering algorithms based on the intuitionistic fuzzy tolerance relations. A hierarchy of allotments among the intuitionistic fuzzy clusters will be the result of classification in this case.

3.2 A Method of Clustering the Fuzzy Data

The first subsection of the section provides a brief consideration of basic approaches to the clustering of fuzzy data. Distances for the fuzzy numbers and a method of computing dissimilarity coefficients between vectors of fuzzy numbers are considered in the second subsection. In the third subsection illustrative examples are shown.

3.2.1 Related Works

Fuzzy data is quite a natural type of data, like non-precise data. A natural consequence is that we face the problem of clustering of fuzzy data. The problem is very relevant, for example, in medical diagnosis and many military applications. Let us consider the problem of fuzzy data clustering in detail.

Yang and Ko [164] proposed a class of fuzzy c -number clustering procedures for fuzzy data clustering which are based on the objective function (1.106). These so-called FCN-algorithms were developed by Yang and Liu [165] for conical fuzzy vector data.The approach was then developed by Hung and Yang [53] for an exponential-type distance function and the AFCN-algorithm was proposed. However, the FCN and AFCN clustering procedures are objective function-based fuzzy clustering algorithms and a fuzzy c -partition is a result of application of the FCN and AFCN clustering procedures to the data set. Moreover, a set of fuzzy numbers is the input data for the FCN and AFCN fuzzy clustering algorithms. In other words, elements of the initial set $X = \{x_1,\ldots,x_n\}$ of n objects are fuzzy numbers. It should be noted that the D-AFC(c)-algorithm can also be used for the clustering of fuzzy numbers. The use of the D-AFC(c)-algorithm for the clustering of fuzzy numbers was proposed in [125].

However, in many practical clustering problems objects are described by a set of attributes with fuzzy values. For example, a modification of the FCM-algorithm for fuzzy attributes was proposed by Butkiewicz and Nieradka [16]. It should be noted that the concept of a vector of fuzzy numbers was introduced in [132] and developed in [136]. The method of processing a set of vectors of fuzzy numbers is based on the application of the D-AFC(c)-algorithm to the corresponding fuzzy tolerance matrix.

3.2.2 Notes on Fuzzy Data Preprocessing

Let us consider distances between fuzzy intervals and fuzzy numbers which were proposed by Yang and Ko [164]. A distance between any two triangular fuzzy numbers is defined by the formula (1.107). So, other distances must be considered.

The set of all LR -type fuzzy intervals will be denoted by $\mathcal{F}_{(LR)FI}(\Re)$ and a set of n fuzzy intervals in $\mathcal{F}_{(LR)FI}(\Re)$ will be denoted by $X_{(LR)FI} = \{V_1,\ldots,V_n\}$. A distance $d^2_{(LR)FI}(V_i,V_j)$ for any $V_i = (\underline{m}_i,\overline{m}_i,a_i,b_i)_{LR}$ and $V_j = (\underline{m}_j,\overline{m}_j,a_j,b_j)_{LR}$ in $\mathcal{F}_{(LR)FI}(\Re)$ is defined as follows:

$$d^2_{(LR)FI}(V_i,V_j) = (\underline{m}_i - \underline{m}_j)^2 + (\overline{m}_i - \overline{m}_j)^2 + \left((\underline{m}_i - la_i) - (\underline{m}_j - la_j)\right)^2 + \left((\overline{m}_i + rb_i) - (\overline{m}_j + rb_j)\right)^2, \quad (3.37)$$

where $l = \int_0^1 L^{-1}(\omega)d\omega$ and $r = \int_0^1 R^{-1}(\omega)d\omega$.

Let $\mathcal{F}_{(T)FI}(\Re)$ be the space of all trapezoidal fuzzy intervals and $X_{(T)FI} = \{V_1,\ldots,V_n\}$ be a set of n trapezoidal fuzzy intervals in $\mathcal{F}_{(T)FI}(\Re)$. According to the definition of a distance $d^2_{(LR)FI}(V_i,V_j)$ as defined before, a distance $d^2_{(T)FI}(V_i,V_j)$ for any two trapezoidal fuzzy intervals $V_i=(\underline{m}_i,\overline{m}_i,a_i,b_i)_{TI}$ and $V_j=(\underline{m}_j,\overline{m}_j,a_j,b_j)_{TI}$ can be defined as follows:

$$d^2_{(T)FI}(V_i,V_j)=(\underline{m}_i-\underline{m}_j)^2+(\overline{m}_i-\overline{m}_j)^2+ \\ +\left((\underline{m}_i-\frac{1}{2}a_i)-(\underline{m}_j-\frac{1}{2}a_j)\right)^2+\left((\overline{m}_i+\frac{1}{2}b_i)-(\overline{m}_j+\frac{1}{2}b_j)\right)^2. \tag{3.38}$$

Let $\mathcal{F}_{(LR)FN}(\Re)$ denote the set of all LR-type fuzzy numbers and $X=\{V_1,\ldots,V_n\}$ be a set of n fuzzy numbers in $\mathcal{F}_{(LR)FN}(\Re)$. A distance $d^2_{(LR)FN}(V_i,V_j)$ for any $V_i=(m_i,a_i,b_i)_{LR}$ and $V_j=(m_j,a_j,b_j)_{LR}$, $V_i,V_j\in\mathcal{F}_{(LR)FN}(\Re)$ can be defined as follows:

$$d^2_{(LR)FN}(V_i,V_j)=(m_i-m_j)^2+\big((m_i-la_i)-(m_j-la_j)\big)^2+\big((m_i+rb_i)-(m_j+rb_j)\big)^2, \tag{3.39}$$

where $l=\int_0^1 L^{-1}(\omega)d\omega$ and $r=\int_0^1 R^{-1}(\omega)d\omega$.

Now let $\mathcal{F}_{(G)FN}(\Re)$ be the set of all Gaussian fuzzy numbers. Let $X_{(G)FN}=\{V_1,\ldots,V_n\}$ be a set of n Gaussian fuzzy numbers from $\mathcal{F}_{(G)FN}(\Re)$. Then, a distance $d^2_{(G)FN}(V_i,V_j)$ for any two Gaussian fuzzy numbers $V_i=(m_i,\sigma_i)_G$ and $V_j=(m_j,\sigma_j)_G$ in $\mathcal{F}_{(G)FN}(\Re)$ is defined as follows:

$$d^2_{(G)FN}(V_i,V_j)=3(m_i-m_j)^2+\frac{\pi}{2}(\sigma_i-\sigma_j)^2. \tag{3.40}$$

In general, a set $X_{(LR)FI}=\{V_1,\ldots,V_n\}$ of LR-type fuzzy intervals, a set $X_{(T)FI}=\{V_1,\ldots,V_n\}$ of trapezoidal fuzzy intervals, a set $X=\{V_1,\ldots,V_n\}$ of LR-type fuzzy numbers, a set $X_{(T)FN}=\{V_1,\ldots,V_n\}$ of triangular fuzzy numbers, and a set $X_{(G)FN}=\{V_1,\ldots,V_n\}$ of Gaussian fuzzy numbers will be generally denoted by $X=\{V_1,\ldots,V_n\}$. Let us consider a method of data preprocessing for the set $X=\{V_1,\ldots,V_n\}$ of fuzzy numbers (fuzzy intervals).

The matrix of coefficients of pair-wise dissimilarities between the objects $\rho_{n\times n}=[\rho_{ij}]$, $i,j=1,\ldots,n$, can be obtained through the application of the corresponding distance to the data set $X=\{V_1,\ldots,V_n\}$. The matrix of pair-wise dissimilarities should be normalized using the formula (2.20) and the fuzzy intolerance matrix $I = [\mu_I(x_i, x_j)]$ is obtained. So, the matrix of the fuzzy tolerance relation $T = [\mu_T(x_i, x_j)]$ should be obtained after the application of the complement (1.26) to matrix $I = [\mu_I(x_i, x_j)]$. It should be noted that the relational objective-function fuzzy clustering procedures can be applied directly to the data given as a matrix of dissimilarity coefficients $\rho_{n\times n} = [\rho_{ij}]$, $i, j = 1, \ldots, n$.

However, the data can be presented as a matrix of attributes $X_{n\times p_1} = [x_i^{t_1}]$, $i = 1, \ldots, n$, $t_1 = 1, \ldots, p_1$, in wich the value $x_i^{t_1}$ is the value of the t_1-th attribute for the i-th object. However, we often have to deal with objects that cannot be described by precise values of attributes. So, a set $X_{(LR)FI} = \{V^1, \ldots, V^{p_1}\}$ of LR-type fuzzy intervals, a set $X_{(T)FI} = \{V^1, \ldots, V^{p_1}\}$ of trapezoidal fuzzy intervals, a set $X_{(LR)FN} = \{V^1, \ldots, V^{p}\}$ of LR-type fuzzy numbers, a set $X_{(T)FN} = \{V^1, \ldots, V^{p_1}\}$ of triangular fuzzy numbers and a set $X_{(G)FN} = \{V^1, \ldots, V^{p_1}\}$ of Gaussian fuzzy numbers can be considered as sets of values of object attributes.

Thus, some fuzzy number or some fuzzy interval $V_i^{t_1}$, $i \in \{1, \ldots, n\}$, $t_1 \in \{1, \ldots, p_1\}$ is the value of the t_1-th fuzzy attribute for the i-th object. In the context of this approach, a concept of a vector of fuzzy numbers can be defined as follows (cf. [132]):

Definition 3.15. *A crisp set* $\overline{V} = \{V^{t_1} \mid t_1 = 1, \ldots, p_1\}$ *of* p_1 *fuzzy numbers of the same type is called a vector of fuzzy numbers.*

The particular type of the vector of fuzzy numbers depends on the particular types of its elements. In other words, fuzzy numbers of the same type can be elements of the vector $\overline{V}$. In particular, if a fuzzy number V^t is a fuzzy number of the LR-type for all $t_1 = 1, \ldots, p_1$, then the vector $\overline{V} = \{V^{t_1} \mid t_1 = 1, \ldots, p_1\}$ is the vector of fuzzy numbers of the LR-type. Notable, definition 3.15 is general definition, because fuzzy intervals of the same type can be elements of the vector $\overline{V}$, as well.

Let us consider an example of the vector of fuzzy numbers. Let $x_i \in X$ be an object in the two-dimensional space and the values of object attributes be represented by triangular fuzzy numbers $V_i^{t_1} = (m_i^{t_1}, a_i^{t_1}, b_i^{t_1})_T$. Thus, some vector of triangular fuzzy numbers $\overline{V}_i = (V_i^1, V_i^2)$ corresponds to each element of the set of objects $x_i \in X$. A possible geometrical representation of the vector of fuzzy numbers is shown in Figure 3.5.

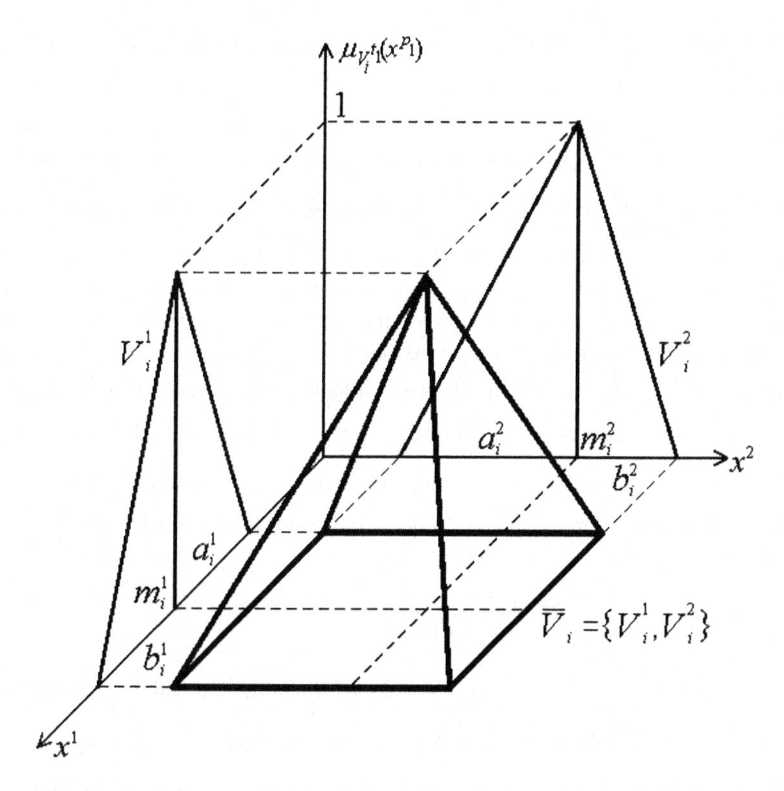

Fig. 3.5 Geometrical representation of a vector of triangular fuzzy numbers in a two-dimensional space

Thus, the vector of fuzzy numbers in some p_1-dimensional space is an area bounded by the corresponding fuzzy numbers. It should be noted that the concept of a vector of fuzzy numbers in some p_1-dimensional space can be defined via the Cartesian product of p_1 universes.

Vectors of fuzzy numbers in the sense of Definition 3.15 can be called homogeneous vectors of fuzzy numbers [136], while the concept of a heterogeneous vector of fuzzy numbers is introduced in [136] as follows:

Definition 3.16 *A crisp set* $\overline{V} = \{V^{t_1} \mid t_1 = 1, \ldots, p_1\}$ *of* p_1 *fuzzy numbers or fuzzy intervals of different types is a heterogeneous vector of fuzzy numbers.*

Thus, trapezoidal fuzzy intervals, triangular fuzzy numbers and Gaussian fuzzy numbers can be elements of some heterogeneous vector of fuzzy numbers.

On the other hand, a concept of a fuzzy vector was considered by Bandemer and Näther [7]. A fuzzy set A on a p_1-dimensional vector space $\Re^{p_1}$ is called a fuzzy vector if A is convex and there exists one point $y \in \Re^{p_1}$ with the membership function equal $\mu_A(y) = 1$. Moreover, the concept of a p_1-dimensional fuzzy vector was considered by Viertl [149]. A p_1-dimensional fuzzy vector A is determined by its so-called vector-characterized function $\mu_A : \Re^{p_1} \to [0,1]$ which is a real function of p_1 real variables $x_1, \ldots, x_{p_1}$ such that the support of A is a bounded set and, $\forall \alpha \in (0,1]$, the α-level A_α of the p_1-dimensional fuzzy vector A is non-empty, bounded and is a finite union of simply connected and closed sets. Obviously, the definitions of the vector of fuzzy numbers differ from one definition to the other.

Let $X = \{x_1, \ldots, x_n\}$ be an initial set of elements and a vector of fuzzy numbers $\overline{V}_i$, $i = 1, \ldots, n$, correspond to each object $x_i \in X$. In general, a distance between different objects $x_i = (V_i^1, \ldots, V_i^{p_1})$ and $x_j = (V_j^1, \ldots, V_j^{p_1})$ can be defined as an average value of the sum of distances between the particular attributes:

$$D(x_i, x_j) = \frac{1}{p_1} \sum_{t_1=1}^{p_1} d^2(V_i^{t_1}, V_j^{t_1}), \tag{3.41}$$

where $d^2(V_i^{t_1}, V_j^{t_1})$ is a distance for two fuzzy numbers which represents the values of the same t_1-th fuzzy attribute for different objects $x_i, x_j \in X$.

The distance (3.41) is a linear combination of p_1 distances between the attributes. So, this distance depends on the type of fuzzy numbers which represent the attributes. For example, if the attributes of an object are represented by the triangular fuzzy numbers, $V_i^{t_1} = (m_i^{t_1}, a_i^{t_1}, b_i^{t_1})_T$, $i = 1, \ldots, n$, $t_1 = 1, \ldots, p_1$, then the distance (3.41) can be rewritten as follows:

$$D(x_i, x_j) = \frac{1}{p_1} \sum_{t_1=1}^{p_1} d^2_{(T)FN}(V_i^{t_1}, V_j^{t_1}), \tag{3.42}$$

where $d^2_{(T)FN}(V_i^{t_1}, V_j^{t_1})$ is the distance for triangular fuzzy numbers (1.107) which is rewritten here as:

$$\begin{aligned} d^2_{(T)FN}(V_i^{t_1}, V_j^{t_1}) = (m_i^{t_1} - m_j^{t_1})^2 + \left((m_i^{t_1} - m_j^{t_1}) - \frac{1}{2}(a_i^{t_1} - a_j^{t_1}) \right)^2 + \\ + \left((m_i^{t_1} - m_j^{t_1}) + \frac{1}{2}(b_i^{t_1} - b_j^{t_1}) \right)^2 \end{aligned}, \tag{3.43}$$

for all $i, j = 1, \ldots, n$, $t_1 = 1, \ldots, p_1$.

By applying the distance (3.41) to the data set $X = \{x_1,..., x_n\}$ a matrix of coefficients of pair-wise dissimilarities between the objects $\rho_{n \times n} = [\rho_{ij}]$, $i, j = 1,\dots,n$, can be obtained. Thus, the fuzzy tolerance matrix $T = [\mu_T(x_i, x_j)]$, $i, j = 1,\dots,n$, can be obtained using the formula (2.20) and the complement (1.26).

The technique of preprocessing of a set of heterogeneous vectors of fuzzy numbers can be described in the similar way. However, the classification of a set of fuzzy numbers and of a set of vectors of fuzzy numbers should be illustrated by examples.

3.2.3 *Illustrative Examples*

An example of the D-AFC(c)-algorithm performance for classification of a set of fuzzy numbers can be illustrated by a set of triangular fuzzy numbers which are shown in Figure 1.3.

The distance (1.107) was applied to the set $X_{(T)FN} = \{V_1,\dots,V_{30}\}$ of triangular fuzzy numbers and the matrix of dissimilarity coefficients $\rho_{30 \times 30} = [\rho_{ij}]$, $i, j = 1,\dots,30$, was obtained. So, the fuzzy tolerance matrix $T = [\mu_T(x_i, x_j)]$, $i, j = 1,\dots,30$, was constructed. By executing the D-AFC(c)-algorithm for three classes we obtain the following: the first class is formed by 4 elements; the second class by 17 elements; the third class by 9 elements. The allotment which corresponds to the result was obtained for the tolerance threshold $\alpha = 0.9345$. The membership functions of three fuzzy clusters are presented in Figure 3.6. The membership values of the first class are represented by ○, the membership values of the second class are represented by ■ and the membership values of the third class are represented by ▲.

The value of the membership function of the fuzzy cluster which corresponds to the first class attains its maximum for the first object and is equal one. So, the fuzzy number $V_1 = (m_1 = 3.34, a_1 = 1.46, b_1 = 1.30)_T$ is the typical point of the fuzzy cluster which corresponds to the first class. The membership value of the fifteenth object is equal one for the fuzzy cluster which corresponds to the second class. Thus, the fuzzy number $V_{15} = (m_{15} = 23.47, a_{15} = 0.81, b_{15} = 0.51)_T$ is the typical point of the fuzzy cluster which corresponds to the second class. The membership value of the thirtieth object is equal one for the fuzzy α-cluster which corresponds to the third class. That is why the fuzzy number $V_{30} = (m_{30} = 45.77, a_{30} = 1.71, b_{30} = 0.79)_T$ is the typical point of the fuzzy cluster which corresponds to the third class.

The result of the numerical experiment seems to be satisfactory in comparison with the result obtained from the FCN-algorithm which is presented in Figure 1.4. Note however that we have preassumed the number $c = 3$ of clusters in the

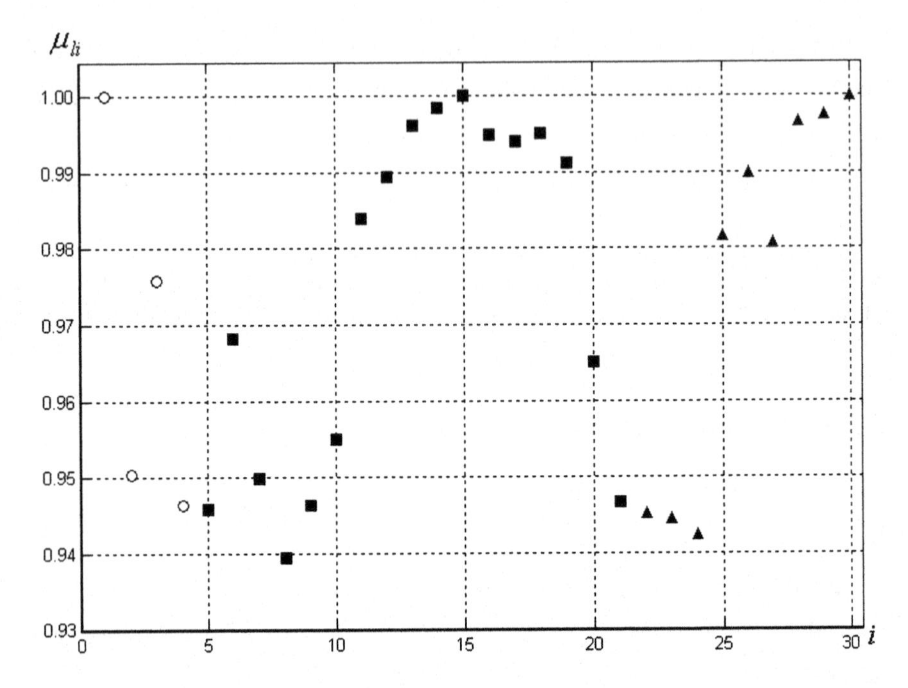

Fig. 3.6 The membership functions of three fuzzy clusters obtained by the D-AFC(c)-algorithm

allotment sought because the this number was determined by Yang and Ko [164] in their numerical experiments. On the other hand, the data was processed by the D-AFC(c)-algorithm with the number of fuzzy clusters $c = 2,\ldots,5$ using the linear measure of fuzziness of the allotment (2.30). The performance of the validity measure is shown in Figure 3.7.

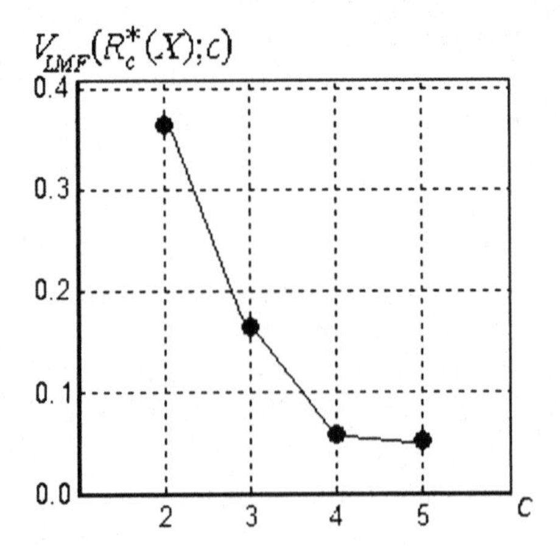

Fig. 3.7 Values of the linear measure of fuzziness of the allotment as a function of the number of clusters

So, the maximal value of the linear measure of fuzziness of the allotment is equal to 0.3647 and this value corresponds to two fully separate fuzzy clusters. The corresponding allotment $R_c^*(X)$ was obtained for the tolerance threshold $\alpha = 0.9726$. By executing the D-AFC(c)-algorithm for two classes we obtain the following: the first class is formed by 13 elements and the second class by 17 elements. The membership functions of two fuzzy clusters are presented in Figure 3.8. The membership values of the first class are represented by ○ and the membership values of the second class are represented by ■.

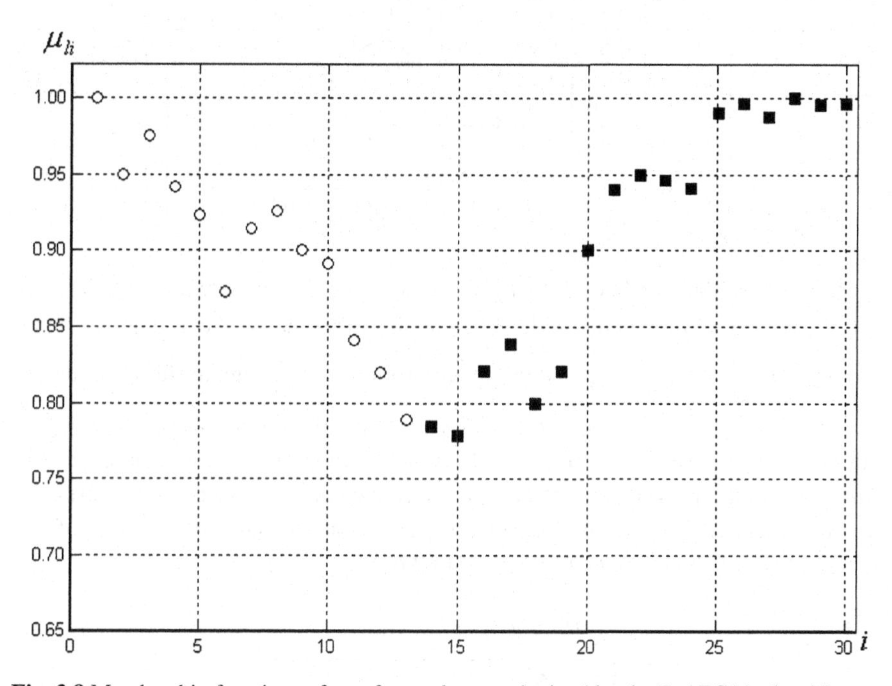

Fig. 3.8 Membership functions of two fuzzy clusters obtained by the D-AFC(c)-algorithm

The fuzzy number $V_1 = (m_1 = 3.34, a_1 = 1.46, b_1 = 1.30)_T$ is the typical point of the fuzzy cluster which corresponds to the first class. The fuzzy number $V_{28} = (m_{28} = 43.56, a_{28} = 0.92, b_{28} = 0.63)_T$ is the typical point of the fuzzy cluster which corresponds to the second class.

Let us consider a problem of classification of a set of vectors of fuzzy numbers. For this purpose, Yang and Ko's [164] data set of thirty triangular fuzzy numbers were modified [132] and is shown in Table 3.5.

Thus, the vector of triangular fuzzy numbers $\overline{V_i} = (V_i^1, V_i^2, V_i^3)$ corresponds to each element of the set of objects $X = \{x_1, \ldots, x_{10}\}$. A fuzzy tolerance matrix $T = [\mu_T(x_i, x_j)]$, $i, j = 1, \ldots, 10$,was obtained by applying the considered technique to the data set.

Table 3.5 Data set for numerical experiments

Numbers of objects, i	Attributes								
	$V^1=(m^1,a^1,b^1)_T$			$V^2=(m^2,a^2,b^2)_T$			$V^3=(m^3,a^3,b^3)_T$		
	m^1	a^1	b^1	m^2	a^2	b^3	m^3	a^3	b^3
1	3.34	1.46	1.30	19.78	1.47	0.42	32.77	0.63	0.47
2	9.56	0.27	1.00	20.67	1.34	1.10	34.88	1.08	0.66
3	10.56	1.95	1.93	21.45	0.92	1.60	35.45	1.48	1.26
4	10.89	0.56	1.17	22.34	0.04	1.58	35.88	1.79	0.16
5	13.89	0.89	0.88	23.47	0.81	0.51	38.88	0.66	0.64
6	14.78	0.12	1.21	24.67	0.14	1.09	40.25	0.52	1.71
7	14.90	1.19	0.41	25.78	0.39	1.51	40.47	1.95	0.15
8	15.67	1.82	0.90	26.45	1.61	0.92	43.56	0.92	0.63
9	16.87	1.90	1.85	28.34	1.95	0.12	43.98	1.74	1.69
10	17.45	1.79	1.95	32.29	1.66	1.64	45.77	1.71	0.79

Intuitively, the number of fuzzy clusters $c=2$ is suitable for the data set $X=\{x_1,\ldots,x_{10}\}$ shown in Table 3.5. On the other hand, the triangular fuzzy number $V_1^1=(m_1^1,a_1^1,b_1^1)_T$ is far away from other triangular fuzzy numbers $V_i^1=(m_i^1,a_i^1,b_i^1)_T$, $i=2,\ldots,10$. That is why the first object can be regarded as an element of a separate class. So, the D-AFC(c)-algorithm was applied to the fuzzy tolerance matrix for $c=2,\ldots,5$ using the linear measure of fuzziness of the allotment (2.30). The performance of the linear measure of fuzziness of the allotment for the data set is presented in Figure 3.9.

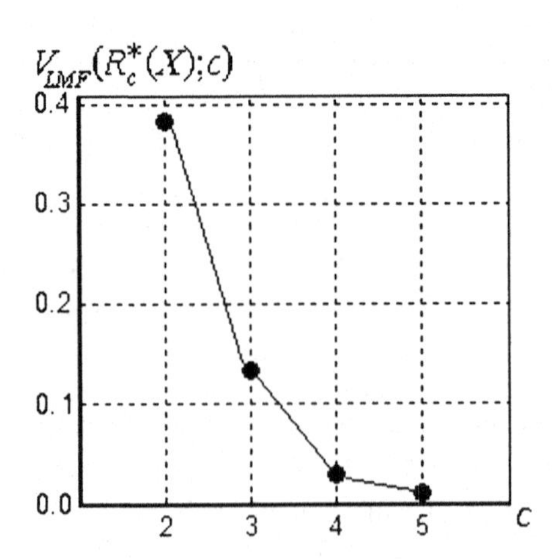

Fig. 3.9 Values of the linear measure of fuzziness of the allotment as a function of the number of clusters

The maximal value of the validity measure is equal to 0.3861 and this value corresponds to two fuzzy clusters.

The allotment $R_c^*(X)$ among two fully separate fuzzy clusters, which corresponds to the result sought, is obtained for the tolerance threshold $\alpha = 0.73707504$. The membership functions of two classes of the allotment are presented in Figure 3.10 and the values which equal zero are not shown in this figure. The membership values of the first class are represented by ○ and the membership values of the second class are represented by ■.

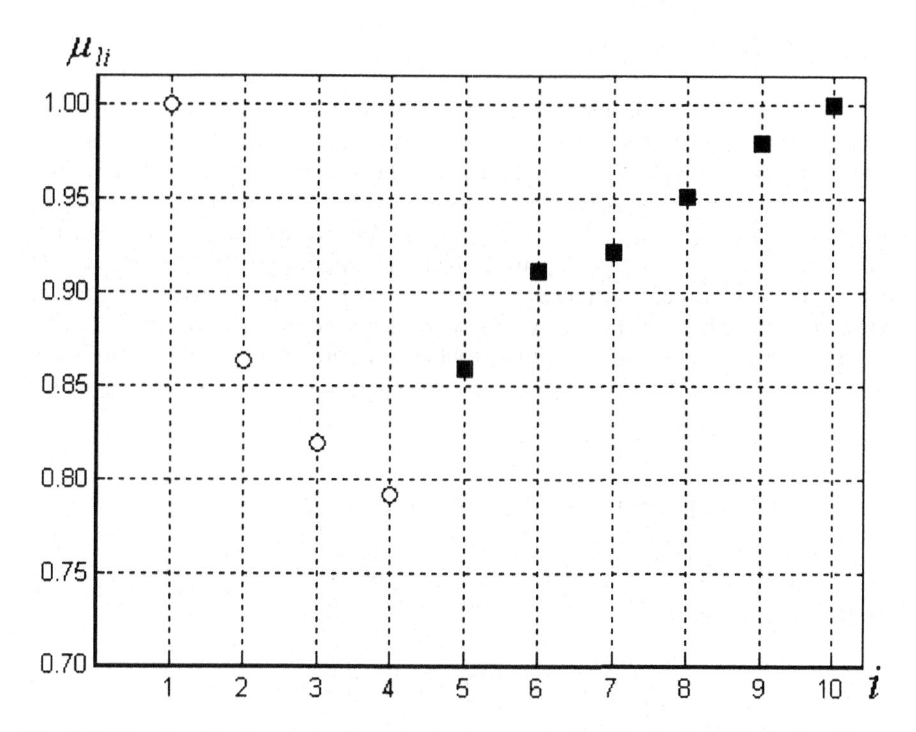

Fig. 3.10 Membership functions of two fuzzy clusters obtained by the D-AFC(c)-algorithm

The value of the membership function of the fuzzy cluster, which corresponds to the first class, is maximal for the first object and is equal one. So, the first object is the typical point of the fuzzy cluster which corresponds to the first class. The membership value of the tenth object is equal one and this value is maximal for the fuzzy cluster which corresponds to the second class. That is why the tenth object is the typical point of the fuzzy cluster which corresponds to the second class.

The results of numerical experiments shows that, on the one hand, the proposed fuzzy data preprocessing methodologies are suitable and, on the other hand, the D-AFC(c)-algorithm is an effective and efficient clustering procedure for the classification of fuzzy data.

3.3 Clustering the Three-Way Data

In the first subsection of the section, a brief review of fuzzy approaches to the clustering of three-way data is provided, in the second subsection a technique for the preprocessing of the three-way data is described, in the third subsection an illustrative example of the Sato and Sato's [100] three-way data preprocessing is shown and experimental results of the application of the D-AFC(c)-algorithm using a validity measure to the data set are given and discussed.

3.3.1 *Preliminary Remarks*

A problem of fuzzy clustering for three-way data was outlined by Sato and Sato [100]. Typical three-way data are composed of objects, attributes and situations, exemplified by the height and weight of children at several ages. The purpose of clustering for three-way data is to reveal the latent structure through all the time related situations by constructing clusters which take into account not only the similarity between the pair of objects at individual time instants but also the similarity between the patterns of change of observation over time. Fuzzy clustering procedure for three-way data was considered in [100] as a procedure of solving a multicriteria optimization problem. In particular, the three-way data, which are observed by the values of m_1 attributes with respect to n objects for m_2 situations, are denoted as:

$$\hat{X}^{(t_2)}_{n \times m_1} = [\hat{x}_i^{t_1(t_2)}],\ i = 1,\ldots,n\,,\ t_1 = 1,\ldots,m_1\,,\ t_2 = 1,\ldots,m_2\,. \tag{3.44}$$

Following the equation (1.72), the quality of clustering in the t_2-th situation is defined as (cf. [100]):

$$Q^{(t_2)}_{FCM}(P, \mathrm{T}^{(t_2)}) = \sum_{l=1}^{c}\sum_{i=1}^{n} u_{li}^{\gamma} d^2(x_i^{(t_2)}, \tau^{l(t_2)})\,, \tag{3.45}$$

where $\tau^{l(t_2)}$ is the centroid of the l-th fuzzy cluster $A^l \in P(X)$ at the t_2-th situation and $d^2(x_i^{(t_2)}, \tau^{l(t_2)})$ is the squared Euclidean distance (1.73) between $x_i^{(t_2)}$ and $\tau^{l(t_2)}$ at the t_2-th situation.

If there exists a solution which minimizes all criteria $Q^{(t_2)}_{FCM}(P, \mathrm{T}^{(t_2)})$, $t_2 = 1,\ldots,m_2$, then it is the best (optimal) solution. Such a solution does not exist very often. Then this problem, a clustering for three-way data, becomes a multicriteria optimization problem. Let us assume that Φ is a set of feasible solutions $P(X)$. A single clustering criterion is defined by the weighted sum of criteria $Q^{(t_2)}_{FCM}(P, \mathrm{T}^{(t_2)})$ (cf. [100])

$$Q(P,\mathrm{T}^{(t_2)}) = \sum_{t_2=1}^{m_2} w^{(t_2)} Q_{FCM}^{(t_2)}(P,\mathrm{T}^{(t_2)}), \tag{3.46}$$

for all $w^{(t_2)} > 0$. The equation (3.46) shows that the problem of finding a Pareto optimal solution of the multicriteria problem $\left(\Phi, Q_{FCM}^{(1)}(P,\mathrm{T}^{(1)}),\ldots,Q_{FCM}^{(m_2)}(P,\mathrm{T}^{(m_2)})\right)$ is reduced to the usual nonlinear optimization problem with one criterion (cf. [100]).

The method of fuzzy clustering of Sato and Sato [100] was extended by Sato-Ilic and Jain [102] for the relational data. The three-way data, which are observed by the values of dissimilarity with respect to n objects for m_2 situations, are denoted as follows (cf. [102]):

$$\rho_{n\times n}^{(t_2)} = [d^{(t_2)}(x_i, x_j)],\ i,j = 1,\ldots,n,\ t_2 = 1,\ldots,m_2. \tag{3.47}$$

So, the purpose of clustering is to classify n objects into c fuzzy clusters which are the fuzzy subsets of the set of objects $X = \{x_1,\ldots,x_n\}$. From the equation (1.82), the quality of clustering in the t_2-th situation is given [102] by a modified sum of values of extended within-class dispersions:

$$Q_{RFCM}^{(t_2)}(P) = \sum_{l=1}^{c}\left(\sum_{j=1}^{n}\sum_{i=1}^{n} u_{li}^{\gamma} u_{lj}^{\gamma} d^{(t_2)}(x_i, x_j) \Big/ 2\sum_{k=1}^{n} u_{lk}^{\gamma}\right). \tag{3.48}$$

The best solution is the solution which minimizes all $Q_{RFCM}^{(t_2)}(P)$, $t_2 = 1,\ldots,m_2$. However, such a solution does not normally exist. A multicriteria optimization problem appears. A single clustering criterion was defined in [102] by the weighted sum of $Q_{FCM}^{(t_2)}(P,\mathrm{T}^{(t_2)})$, that is, for all weights $w^{(t_2)} > 0$,

$$Q(P) = \sum_{t_2=1}^{m_2} w^{(t_2)} Q_{RFCM}^{(t_2)}(P). \tag{3.49}$$

and herefore the problem of finding a Pareto optimal solution is reduced to the solution of an ordinary nonlinear optimization problem. The fuzzy c-partition $P(X)$ in the sense of (1.71) is the clustering result obtained from both methods.

Another relational fuzzy c-means for thee three-way data was also proposed in [102] and that method was referred to a dynamic relational fuzzy c-means. An objective function for that method and an outline of the corresponding DRFCM-algorithm are considered in [102] in detail.

On the other hand, very interesting results for three-way data were described by Coppi and D'Urso [23], [24]. In particular, fuzzy multivariate time trajectories are defined, three types of dissimilarity measures are introduced and three corresponding types of dynamic fuzzy clustering models are suggested. Those models are based on a generalization of Yang and Ko's [164] objective functions for fuzzy clustering.

3.3.2 A Technique of the Three-Way Data Preprocessing

The problem of clustering the three-way data can be formulated as follows [134]. Let $X = \{x_1,...,x_n\}$ be a set of objects, where objects are indexed by i, $i = 1,\ldots,n$; each object x_i is described by m_1 attributes, indexed t^1, $t^1 = 1,\ldots,m_1$, so that an object x_i can be represented by a vector $x_i = (x_i^1,\ldots,x_i^{t_1},\ldots,x_i^{m_1})$; each attribute $\hat{x}^{t_1}$, $t_1 = 1,\ldots,m_1$, can be characterized by m_2 values of 2-ary attributes so that $\hat{x}_i^{t_1} = (\hat{x}_i^{t_1(1)},\ldots,\hat{x}_i^{t_1(t_2)},\ldots,\hat{x}_i^{t_1(m_2)})$.

The three-way data can therefore be presented by a poly-matrix:

$$\hat{X}_{n \times m_1 \times m_2} = [\hat{x}_i^{t_1(t_2)}], \; i = 1,\ldots,n, \; t_1 = 1,\ldots,m_1, \; t_2 = 1,\ldots,m_2. \tag{3.50}$$

The three-way data are the data which are observed by the values of m_1 attributes with respect to n objects for m_2 situations. The purpose of the clustering is to classify the set $X = \{x_1,...,x_n\}$ into c fuzzy clusters and the number of clusters c can be unknown because it depends on the situation. In other words, the cluster structure of the set of objects $X = \{x_1,..., x_n\}$ must be stable at each situation.

The D-AFC(c)-algorithm can be applied directly to the data given as fuzzy tolerance matrix $T = [\mu_T(x_i, x_j)]$, $i, j = 1,\ldots,n$. This means that it can be used by choosing a suitable metric to measure the similarity.

However, the three-way data can be normalized. In particular, the formula (2.11) can be generalized as follows:

$$x_i^{t_1(t_2)} = \frac{\hat{x}_i^{t_1(t_2)}}{\max\limits_{i,t_2} \hat{x}_i^{t_1(t_2)}}. \tag{3.51}$$

On the other hand, the three-way data can be normalized using a generalization of the formula (2.12), which can be written as follows:

$$x_i^{t_1(t_2)} = \frac{\hat{x}_i^{t_1(t_2)} - \min\limits_{i,t_2} \hat{x}_i^{t_1(t_2)}}{\max\limits_{i,t_2} \hat{x}_i^{t_1(t_2)} - \min\limits_{i,t_2} \hat{x}_i^{t_1(t_2)}}. \tag{3.52}$$

So, each object $x_i, i = 1,\ldots,n$, from the initial set $X = \{x_1,\ldots,x_n\}$ can be considered as a type-two fuzzy set and $x_i^{t_1(t_2)} = \mu_{x_i}(x^{t_1(t_2)})$, $i = 1,\ldots,n$;

$t_1 = 1, \ldots, m_1$, $t_2 = 1, \ldots, m_2$, $x^{t_1(t_2)} = \mu_{t_1}(x^{t_2}) \in [0,1]$, $t_1 = 1, \ldots, m_1$, $t_2 = 1, \ldots, m_2$ are its membership functions. In the case of three-way data each object x_i, $i = 1, \ldots, n$, can be presented as the matrix $X_{(i)m_1 \times m_2} = [x_i^{t_1(t_2)}]$, $t_1 = 1, \ldots, m_1$, $t_2 = 1, \ldots, m_2$.

The concept of a type-two fuzzy set was introduced by Zadeh [171] as an extension of the concept of an ordinary fuzzy set, which was called a type-one fuzzy set. The advances of type-two fuzzy sets for pattern recognition were considered by Zeng and Liu [173].

Dissimilarity coefficients between the objects represented by type-two fuzzy sets can be constructed on the basis of generalizations of distances between the fuzzy sets (2.13) – (2.15) and these generalizations can take into account dissimilarities between the object attributes as well as the attribute situations. So, many generalizations of the distances for fuzzy sets can be introduced as functions of dissimilarities – cf. [121] and [134].

The corresponding functions of dissimilarities can be written as follows:

- A generalization of the normalized Hamming distance for the three-way data:

$$l_G(x_i, x_j) = \frac{1}{m_1} \sum_{t_1=1}^{m_1} \left(\frac{1}{m_2^2} \sum_{u_1, v_1=1}^{m_2} | \mu_{x_i}(x^{t_1(u_1)}) - \mu_{x_j}(x^{t_1(v_1)}) | \right), i, j = 1, \ldots, n, \quad (3.53)$$

- A generalization of the normalized Euclidean distance for the three-way data:

$$e_G(x_i, x_j) = \sqrt{\frac{1}{m_1} \sum_{t_1=1}^{m_1} \left(\frac{1}{m_2^2} \sum_{u_1, v_1=1}^{m_2} \left(\mu_{x_i}(x^{t_1(u_1)}) - \mu_{x_j}(x^{t_1(v_1)}) \right)^2 \right)}, i, j = 1, \ldots, n, \quad (3.54)$$

- A generalization of the squared normalized Euclidean distance for the three-way data:

$$\varepsilon_G(x_i, x_j) = \frac{1}{m_1} \sum_{t_1=1}^{m_1} \left(\frac{1}{m_2^2} \sum_{u_1, v_1=1}^{m_2} \left(\mu_{x_i}(x^{t_1(u_1)}) - \mu_{x_j}(x^{t_1(v_1)}) \right)^2 \right), i, j = 1, \ldots, n. \quad (3.55)$$

Obviously, for $m_2 = 1$ the usual distance for the fuzzy sets will be obtained in each case. Properties of the functions of dissimilarities (3.53) – (3.55) are a subject of a special study.

A value m_2 can be different for different attributes $\hat{x}^{t_1}$, $t_1 \in \{1, \ldots, m_1\}$, or a value m_2 of grades for a fixed attribute $\hat{x}^{t_1}$, $t_1 \in \{1, \ldots, m_1\}$ can be different for different objects x_i, $i \in \{1,\ldots,n\}$. So, each object x_i, $i = 1,\ldots,n$, cannot be presented as a matrix $X_{(i)m_1 \times m_2} = [x_i^{t_1(t_2)}]$, $t_1 = 1,\ldots,m_1$, $t_2 = 1,\ldots,m_2$, because a value m_2, which is general for all attributes $\hat{x}^{t_1}$, $t_1 \in \{1,\ldots,m_1\}$, must be established. In these cases a value m_2 can be generally defined as follows:

$$m_2 = \max_{t_1} m_2^{(t_1)}, \; t_1 = 1,\ldots,m_1, \tag{3.56}$$

where the number of grades of every attribute $\hat{x}^{t_1}, t_1 \in \{1,\ldots,m_1\}$ is denoted by $m_2^{(t_1)}$. However, values $x_i^{t_1(t_2)}, i \in \{1,\ldots,n\}$ may not be known for some objects $x_i \in X, i \in \{1,\ldots,n\}$. In such a case, the unknown values $x_i^{t_1(t_2)}, i \in \{1,\ldots,n\}$ can be defined as follows:

$$x_i^{t_1(t_2)} = \max_{t_1} t_2^{(t_1)}, \; i \in \{1,\ldots,n\}, \; t_2 = 1,\ldots,m_2^{(t_1)}. \tag{3.57}$$

Therefore, the method of the three-way data preprocessing can be very simply generalized for the case of p-way data.

3.3.3 An Illustrative Example

The method described will now be explained by an example. For this purpose, Sato and Sato's [100] three-way data set is used. Their artificial three-way data come from a medical survey of the body, involving the height, weight, chest girth and sitting height, which constitute the measurements of 38 boys at three time instants, that is, when the individuas in question were 13, 14 and 15 years old. These data have originally appeared in Sato and Sato [100]. That original data can be rewritten as presented in [134].

Denote the height by $\hat{x}^1$, weight by $\hat{x}^2$, chest girth by $\hat{x}^3$ and sitting height by $\hat{x}^4$. Each attribute $\hat{x}^{t_1}$, $t_1 = 1,\ldots,4$, is observed at three instants $t_2 = 1,\ldots,3$. The value of the t_1-th attribute at the t_2-th moment for the i-th object will be denoted by $\hat{x}_i^{t_1(t_2)}$, $i = 1,\ldots,38$, $t_1 = 1,\ldots,4$, $t_2 = 1,\ldots,3$. The methodology of the three-way data preprocessing can be applied directly to this data.The data can be normalized using the formula (3.51) or formula (3.52). For example, the thirteenth object after normalization (3.51) will be presented as the matrix $X_{4\times 3} = [x_{13}^{t_1(t_2)}]$, $t_1 = 1,\ldots,4$, $t_2 = 1,\ldots,3$, as shown in Table 3.6.

Table 3.6 Description of an object as a matrix after the normalization by using the formula (3.51)

Number i of an object	Number of the situation t_2 for an attribute $x^{t_1(t_2)}$	Attributes $x^{t_1(t_2)}$ of the object x_i			
		$x^{1(t_2)}$	$x^{2(t_2)}$	$x^{3(t_2)}$	$x^{4(t_2)}$
13	1	0.8457	0.4659	0.7059	0.8211
	2	0.9029	0.5341	0.7549	0.8632
	3	0.9371	0.5795	0.7941	0.8947

The matrix can be presented as a membership function of the type-two fuzzy set on the attributes and the values of each attribute can be described by a type-one fuzzy set. The membership function of a type-two fuzzy set which describes the thirteenth object and the membership function of a type-one fuzzy set which describes the first attribute of the thirteenth object are presented in Figure 3.11 and Figure 3.12.

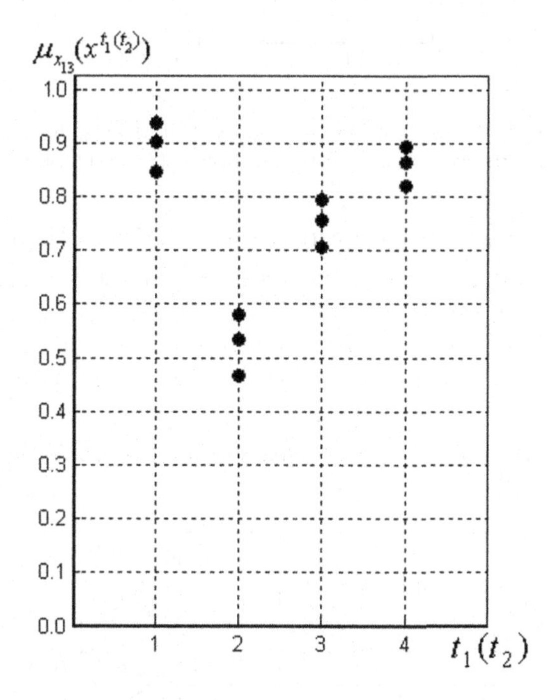

Fig. 3.11 Membership function of the type-two fuzzy set which describes the thirteenth object after the normalization using the formula (3.51)

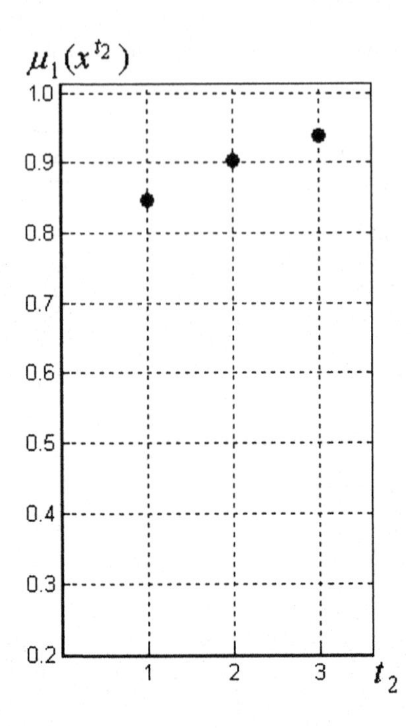

Fig. 3.12 Membership function of the type-one fuzzy set describing the first attribute of the thirteenth object after the normalization using the formula (3.51) at three time instants

The thirteenth object after the normalization (3.52) can be presented as the matrix $X_{4\times 3} = [x_{13}^{t_1(t_2)}]$, $t_1 = 1,\ldots,4$, $t_2 = 1,\ldots,3$, shown in Table 3.7. So, the membership function of a type-two fuzzy set which describes the thirteenth object and the membership function of a type-one fuzzy set which describes the first attribute of the thirteenth object are presented in Figures 3.13 and 3.14.

Table 3.7 Description of an object as a matrix after the normalization using the formula (3.52)

A number i of an object	A number of the situation t_2 of an attribute $x^{t_1(t_2)}$	Attributes $x^{t_1(t_2)}$ of the object x_i			
		$x^{1(t_2)}$	$x^{2(t_2)}$	$x^{3(t_2)}$	$x^{4(t_2)}$
13	1	0.2286	0.1897	0.2857	0.2273
	2	0.5143	0.2931	0.4048	0.4091
	3	0.6857	0.3621	0.5000	0.5455

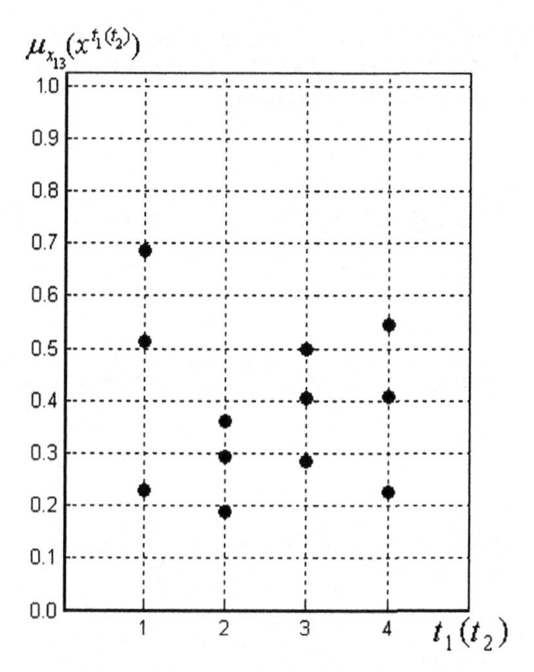

Fig. 3.13 Membership function of the type-two fuzzy set which describes the thirteenth object after normalization using the formula (3.52)

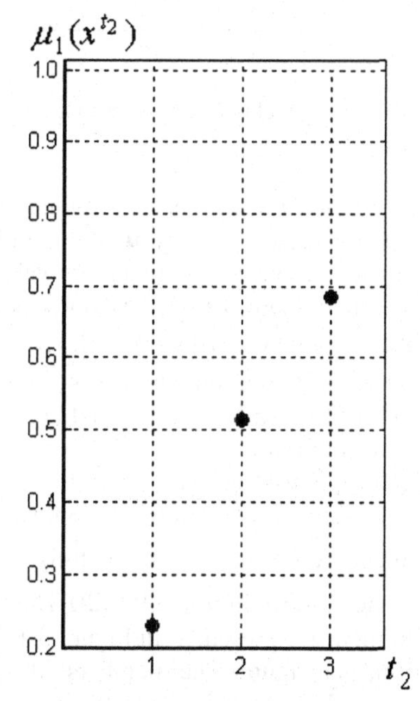

Fig. 3.14 Membership function of the type-one fuzzy set describing the first attribute of the thirteenth object after the normalization using the formula (3.52) at three time instants

Let us consider the classification result obtained from the model (3.45)– (3.46) as was reported by Sato and Sato [100]. The membership functions of four classes of the fuzzy c -partition are presented in Figure 3.15.

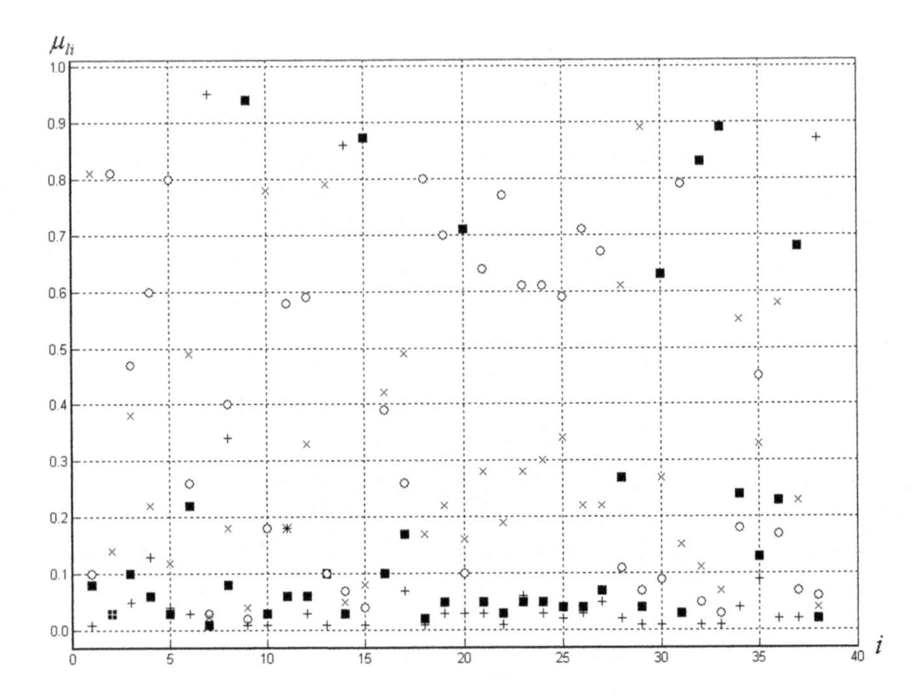

Fig. 3.15 Membership function of four classes obtained by Sato and Sato's classification method

The membership values of the first class are represented by +, the membership values of the second class are represented by ■, the membership values of the third class are represented by ×, and the membership values of the fourth class are represented by ○. For comparison, the data was processed by the D-AFC(c)-algorithm with the number of fuzzy clusters $c = 2,3\ldots$ using the measure of separation and compactness of the allotment given in (2.32). The data was preprocessed according to formulae (3.52), (3.55), and (1.26). The performance of the validity measure is shown in Figure 3.16.

The optimal number of fuzzy clusters is equal 3 and this number corresponds to the first minimum of the measure of separation and compactness of the allotment. The corresponding allotment $R_c^*(X)$ among three fully separate fuzzy clusters was obtained for the tolerance threshold $\alpha = 0.93120$.The membership functions of three classes of the allotment are presented in Figure 3.17 and the values which equal zero are not shown in that figure. The membership values of the first class are represented by +, the membership values of the second class are represented by ■, and the membership values of the third class are represented by ×.

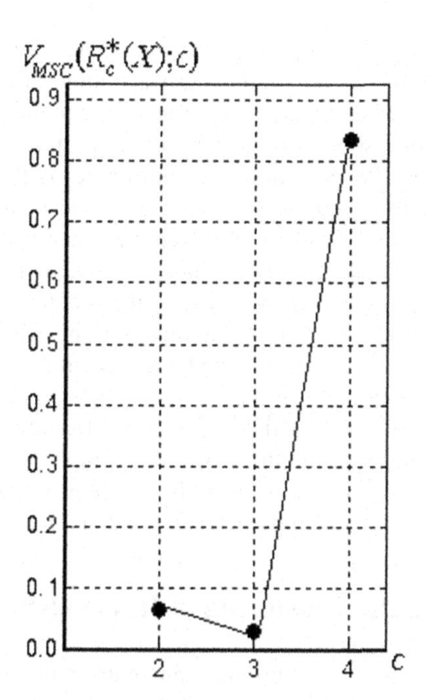

Fig. 3.16 Values of the measure of separation and compactness for Sato and Sato's three-way data set

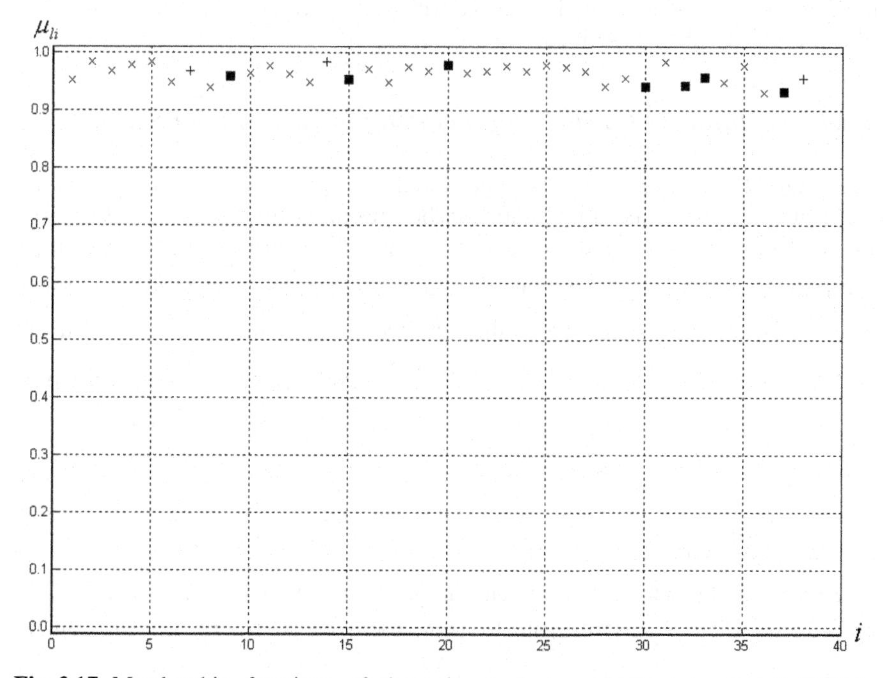

Fig. 3.17 Membership functions of three fuzzy clusters obtained by the D-AFC(c)-algorithm

By executing the D-AFC(c)-algorithm for three classes, we obtain that the first class is formed by 3 elements, the second class is composed of 7 elements, and the third class is formed by 28 elements. The value of the membership function of the fuzzy cluster, which corresponds to the first class, is maximal for the fourteenth object and is equal 0.98298. So, the fourteenth object is the typical point of the fuzzy cluster which corresponds to the first class. The membership value of the twentieth object is equal 0.97888 and this value is maximal for the fuzzy cluster which corresponds to the second class. Thus, the twentieth object is the typical point of the fuzzy cluster which corresponds to the second class. The membership function of the third fuzzy cluster is maximal for the fifth object and is equal 0.98392. That is why the fifth object is the typical point of the fuzzy cluster which corresponds to the third class.

We can notice that the results which are obtained from the D-AFC(c)-algorithm using the described method of the three-way data preprocessing are similar to the results obtained by Sato and Sato [100] using their multicriteria optimization method. However, the proposed approach is more general and simpler.

3.4 A Method for the Clustering of Interval-Valued Data

The first subsection of the section provides an analysis of basic techniques for the preprocessing of the interval-valued data. Distances for interval-valued fuzzy sets and a method of computing dissimilarity coefficients between the vectors of fuzzy numbers are considered in the second subsection. In the third subsection an illustrative example is described.

3.4.1 Techniques for the Preprocessing of Interval-Valued Data

The interval-valued data can be considered as a particular case of the three-way data. The interval-valued data can be presented by a poly-matrix (3.50). In the situation of interval uncertainty, the only information that we have about the actual value $\hat{x}_i^{t_1}$ of some attribute $\hat{x}^{t_1}$, $t_1 \in \{1,\ldots,m_1\}$, for the object x_i, $i \in \{1,\ldots,n\}$ is that the value belongs to some interval, and if $t_2 \in \{\min,\max\}$, then $\hat{x}_i^{t_1} \in [\hat{x}_i^{t_1(\min)}, \hat{x}_i^{t_1(\max)}]$. So, $m_2 = 2$, and the situation can be described by the expression $\hat{x}_i^{t_1} = (\hat{x}_i^{t_1(\min)}, \hat{x}_i^{t_1(\max)})$. Obviously, if $\hat{x}^{t_1(1)} = \ldots = \hat{x}^{t_1(t_2)} = \ldots = \hat{x}^{t_1(m_2)}$, for all $t_1 = 1,\ldots,m_1$, $\forall i = 1,\ldots,n$, then the initial data are the ordinary object data and can be presented as the usual matrix of attributes, $\hat{X}_{n \times m_1} = [\hat{x}_i^{t_1}]$.

That is why the interval-valued data can be normalized using the formulae (3.51) and (3.52), where $t_2 \in \{\min,\max\}$. Thus, each object x_i, $i = 1,\ldots,n$, can be considered as an interval-valued fuzzy set and $\mu_{x_i}(x^{t_1}) = [\mu_{x_i}(x^{t_1(\min)}), \mu_{x_i}(x^{t_1(\max)})]$, $i = 1,\ldots,n$, $t = 1,\ldots,m$, is its membership function, where $\mu_{x_i}(x^{t_1(\min)}) \in [0,1]$, $\mu_{x_i}(x^{t_1(\max)}) \in [0,1]$.

Different distances and similarity measures for the interval-valued fuzzy sets were proposed by different authors. For example, a similarity measure was defined by Ju and Yuan [57] as follows:

$$s_I(x_i, x_j) = 1 - \frac{1}{\sqrt[\lambda]{m_1}} \sqrt[\lambda]{\sum_{t_1=1}^{m_1} \left| \frac{\mu_{x_i}(x^{t_1(\min)}) + \mu_{x_i}(x^{t_1(\max)})}{2} - \frac{\mu_{x_j}(x^{t_1(\min)}) + \mu_{x_j}(x^{t_1(\max)})}{2} \right|^{\lambda}}, \; i, j = 1,\ldots,n, \quad (3.58)$$

where $1 \le \lambda < \infty$. The fuzzy tolerance matrix $T = [\mu_T(x_i, x_j)]$, $i, j = 1,\ldots,n$, is obtained after the application of the similarity measure (3.58) to the set of interval-valued fuzzy sets.

The normalized Euclidean distance between interval-valued fuzzy sets based on the Hausdorff metric was defined by Grzegorzewski [43] as follows:

$$e_I(x_i, x_j) = \sqrt{\frac{1}{m_1} \sum_{t_1=1}^{m_1} \max \left\{ \begin{array}{l} \left(\mu_{x_i}(x^{t_1(\min)}) - \mu_{x_j}(x^{t_1(\min)})\right)^2, \\ \left(\mu_{x_i}(x^{t_1(\max)}) - \mu_{x_j}(x^{t_1(\max)})\right)^2 \end{array} \right\}}, \; i, j = 1,\ldots,n. \quad (3.59)$$

On the other hand, functions of dissimilarities (3.53) – (3.55) can be rewritten for the interval-valued fuzzy sets as follows:

- A generalization of the normalized Hamming distance:

$$l_G(x_i, x_j) = \frac{1}{m_1} \sum_{t_1=1}^{m_1} \left(\frac{1}{2^2} \sum_{t_2 \in \{\min, \max\}} \left| \mu_{x_i}(x^{t_1(t_2)}) - \mu_{x_j}(x^{t_1(t_2)}) \right| \right), i, j = 1,\ldots,n, \quad (3.60)$$

- A generalization of the normalized Euclidean distance:

$$e_G(x_i, x_j) = \sqrt{\frac{1}{m_1} \sum_{t_1=1}^{m_1} \left(\frac{1}{2^2} \sum_{t_2 \in \{\min, \max\}} \left(\mu_{x_i}(x^{t_1(t_2)}) - \mu_{x_j}(x^{t_1(t_2)})\right)^2 \right)}, i, j = 1,\ldots,n, \quad (3.61)$$

- A generalization of the squared normalized Euclidean distance:

$$\varepsilon_I(x_i,x_j)=\frac{1}{m_1}\sum_{t_1=1}^{m_1}\left(\frac{1}{2^2}\sum_{t_2\in\{\min,\max\}}\left(\mu_{x_i}(x^{t_1(t_2)})-\mu_{x_j}(x^{t_1(t_2)})\right)^2\right)\ i,j=1,\ldots,n. \quad (3.62)$$

The distance (3.59) and the functions of dissimilarities (3.60) – (3.62) can be used for constructing the fuzzy intolerance matrix $I=[\mu_I(x_i,x_j)]$, $i,j=1,\ldots,n$. The fuzzy tolerance matrix $T=[\mu_T(x_i,x_j)]$, $i,j=1,\ldots,n$, can be obtained after the application of the complement operation (1.26) to the fuzzy intolerance matrix $I=[\mu_I(x_i,x_j)]$, $i,j=1,\ldots,n$.

3.4.2 Construction of the Stationary Cluster Structure

The initial data matrix (3.50) can be represented as a set of m_2 matrices $\hat{X}^{t_2}_{n\times m_1}=[\hat{x}_i^{t_1}]$, $i=1,\ldots,n$, $t_1=1,\ldots,m_1$, and a "plausible" number c of fuzzy clusters can be different for each matrix $\hat{X}^{t_2}_{n\times m_1}=[\hat{x}_i^{t_1}]$, $t_2\in\{1,\ldots,m_2\}$. A clustering structure of the data set depends on the type of the initial data.

Three types of the clustering structure were defined in [137]. Firstly, if the number of clusters c is some constant for each matrix $\hat{X}^{t_2}_{n\times m_1}=[\hat{x}_i^{t_1}]$, $t_2\in\{1,\ldots,m_2\}$, and the coordinates of the prototypes $\{\tau^1,\ldots,\tau^c\}$ of the clusters $\{A^1,\ldots,A^c\}$ are constant, then the clustering structure is called stable. Secondly, if the actual number of clusters c is some constant for each matrix $\hat{X}^{t_2}_{n\times m_1}=[\hat{x}_i^{t_1}]$, $t_2\in\{1,\ldots,m_2\}$ and the coordinates of prototypes of the clusters are not constant, then the clustering structure is called quasi-stable. Thirdly, if the number of clusters c is different for matrices $\hat{X}^{t_2}_{n\times m_1}=[\hat{x}_i^{t_1}]$, $t_2=1,\ldots,m_2$, then the clustering structure is called unstable.

A method for discovering this type of the clustering structure was proposed in [137] and it can be described as follows: the set of m_2 matrices $X^{t_2}_{n\times m_1}=[x_i^{t_1}]$ can be constructed after the normalization of the initial data matrix $\hat{X}_{n\times m_1\times m_2}=[\hat{x}_i^{t_1(t_2)}]$ and each matrix should be processed by the D-AFC-TC-algorithm by choosing a suitable distance $d(x_i,x_j)$; the number $c^{(t_2)}$ of fully separated fuzzy clusters in the allotment $R_c^*(X^{t_2})$ and the corresponding value of the tolerance threshold $\alpha^{(t_2)}$ are main results of classification. The definition of this type of the clustering structure can be made on the basis of the formulated rules.

The method can be illustrated by the example given by Sato and Jain's [102]. The interval-valued data shown in Table 3.8.

Table 3.8 Sato and Jain's [102] artificial interval-valued data set

Objects	Attributes		
	$\hat{x}^1$	$\hat{x}^2$	$\hat{x}^3$
x_1	[10, 10]	[1, 1]	[2, 2]
x_2	[9, 9]	[3, 3]	[4, 4]
x_3	[13, 13]	[3, 3]	[2, 2]
x_4	[14, 14]	[4, 4]	[5, 5]
x_5	[4, 8]	[11, 11]	[2, 12]
x_6	[6, 10]	[9, 9]	[1, 8]
x_7	[2, 11]	[10, 10]	[1, 11]
x_8	[3, 9]	[8, 8]	[2, 9]

It should be noted that the objects x_1, x_2, x_3, and x_4 have exactly the same values for the minimum and maximum. The matrix of the initial data $\hat{X}_{8\times3\times2} = [\hat{x}_i^{t_1(t_2)}]$, $i=1,\ldots,8$, $t_1=1,\ldots,3$, $t_2 \in \{\min, \max\}$, was normalized using the the formulas (3.51) and (3.52). The two data matrices $X_{8\times3}^{\min} = [x_i^{t_1}]$ and $X_{8\times3}^{\max} = [x_i^{t_1}]$, $i=1,\ldots,8$, $t_1=1,\ldots,3$, were created for each case. For example, the membership functions of an interval-valued fuzzy set which describes the eighth object after corresponding normalizations are presented in Figures 3.18 and 3.19.

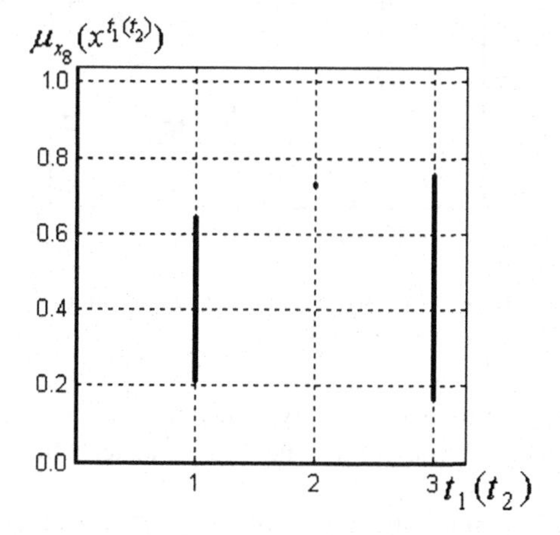

Fig. 3.18 Membership function of the interval-valued fuzzy set which describes the eighth object after the normalization using the formula (3.51)

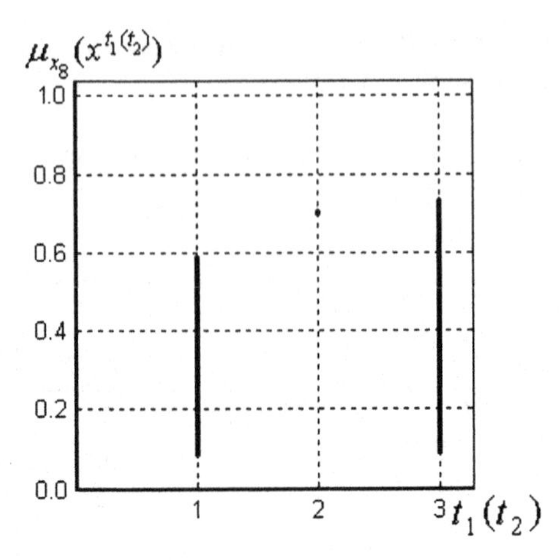

Fig. 3.19 Membership function of the interval-valued fuzzy set which describes the eighth object after the normalization using the formula (3.52)

The two data matrices $X_{8\times3}^{\min}=[x_i^{t_1}]$ and $X_{8\times3}^{\max}=[x_i^{t_1}]$ were constructed after the normalization using the formula (3.51) and both matrices were processed by the D-AFC-TC-algorithm using the normalized Euclidean distance (2.14). Firstly, let us consider the result of the numerical experiment for the data matrix $X_{8\times3}^{\min}=[x_i^{t_1}]$. The allotment $R_c^*(X)$ among two fully separated fuzzy clusters was obtained for the tolerance threshold $\alpha=0.6853$. The membership functions of two classes of the allotment are presented in Figure 3.20 in which the membership values of the first class are represented by ○ and the membership values of the second class are represented by ■.

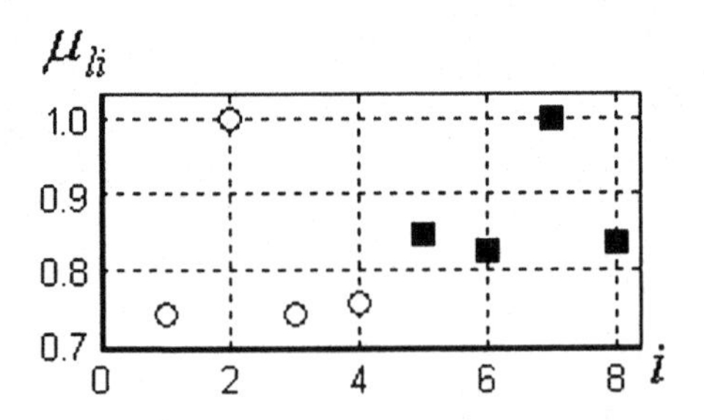

Fig. 3.20 Membership degrees of two fuzzy clusters obtained from the D-AFC-TC-algorithm for the matix of minimal values

Therefore, the second object is the typical point of the first fuzzy cluster and the seventh object is the typical point of the fuzzy cluster which corresponds to the second class.

Secondly, let us consider the result of the experiment for the data matrix $X_{8\times3}^{\max}=[x_i^{t_1}]$. The allotment $R_c^*(X)$ among two fully separated fuzzy clusters was obtained for the tolerance threshold $\alpha = 0.8151$. The membership functions of two classes of the allotment are presented in Figure 3.21. The membership values of the first class are represented by ○ and the membership values of the second class are represented by ■.

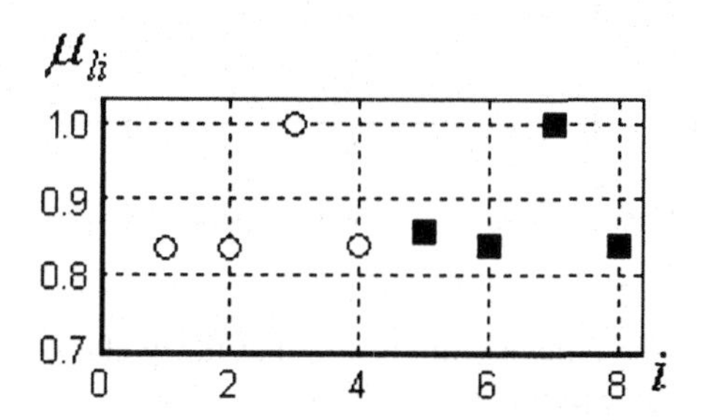

Fig. 3.21 Membership degrees of two fuzzy clusters obtained from the D-AFC-TC-algorithm for the matix of maximal values

The third object is the typical point of the first fuzzy cluster and the seventh object is the typical point of the second fuzzy cluster. Values of the prototypes of fuzzy clusters in both experiments are represented in Table 3.9.

Table 3.9 Prototypes of clusters for both allotments

Class	Result for the matrix of minimal values			Result for the matrix of maximal values		
	Attributes			Attributes		
	$\hat{x}^1$	$\hat{x}^2$	$\hat{x}^3$	$\hat{x}^1$	$\hat{x}^2$	$\hat{x}^3$
1	11.5	2.75	3.25	11.5	2.75	3.25
2	3.75	9.5	1.5	9.5	9.5	10

The current number of clusters c is constant for each matrix $X^{t_2}_{8\times 3} = [x^{t_1}_i]$, $t_2 \in \{\min, \max\}$, and equal 2. The coordinates of prototypes of the clusters are not constant. So, the clustering structure of Sato and Jain's interval-valued data set is the quasi-stable clustering structure.

Let us consider the results of experiments presented in [139]. The formula (3.51) and the similarity measure (3.58) were used in the first experiment. By executing the D-AFC(c)-algorithm for two classes we obtained the following: the first class is formed by 4 elements and the second class is composed of 4 elements. The allotment $R^*_c(X)$ corresponding to the result was obtained for the tolerance threshold $\alpha = 0.57507$ .The membership functions of two classes of the allotment are presented in Figure 3.22. The membership values of the first class are represented by ○ and membership values of the second class are represented by ■.

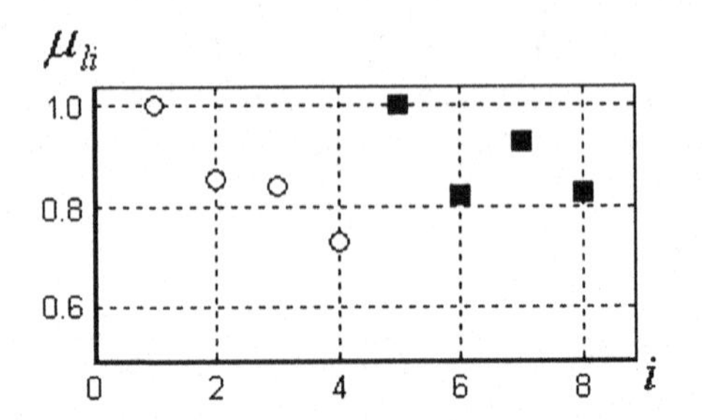

Fig. 3.22 Membership functions of two classes obtained from the D-AFC(c)-algorithm using the formula (3.58)

The first object is the typical point of the first fuzzy cluster and the fifth object is the typical point of the second fuzzy cluster.

The formula (3.51) and the function of dissimilarity (3.61) were used in the second experiment. The matrix of the corresponding feeble fuzzy intolerance relation I_1 is presented in Table 3.10.

Table 3.10 Matrix of dissimilarity coefficients

I_1	x_1	x_2	x_3	x_4	x_5	x_6	x_7	x_8
x_1	0.00000							
x_2	0.14875	0.00000						
x_3	0.16225	0.19097	0.00000					
x_4	0.26988	0.21814	0.15902	0.00000				
x_5	0.65210	0.52642	0.61821	0.56370	0.35964			
x_6	0.48242	0.36961	0.43739	0.40722	0.36348	0.26517		
x_7	0.61099	0.49717	0.57981	0.53801	0.40308	0.37317	0.42967	
x_8	0.48395	0.36475	0.47352	0.44402	0.37193	0.30115	0.38470	0.29550

It is should be noted that the condition (2.19) is not met for the matrix of dissimilarity coefficients. The matrix of the corresponding feeble fuzzy tolerance $T_1 = [\mu_{T_1}(x_i, x_j)]$, $i, j = 1,\ldots,8$, was obtained after the application of the complement operation (1.26) to the matrix of fuzzy intolerance $I_1 = [\mu_{I_1}(x_i, x_j)]$, $i, j = 1,\ldots,8$.

By executing the D-AFC(c)-algorithm, the allotment $R_c^*(X)$ among two fully separated fuzzy clusters was obtained for the tolerance threshold $\alpha = 0.52647$. The membership functions of two classes of the allotment are presented in Figure 3.23 in which the membership values of the first class are represented by ○ and the membership values of the second class are represented by ■.

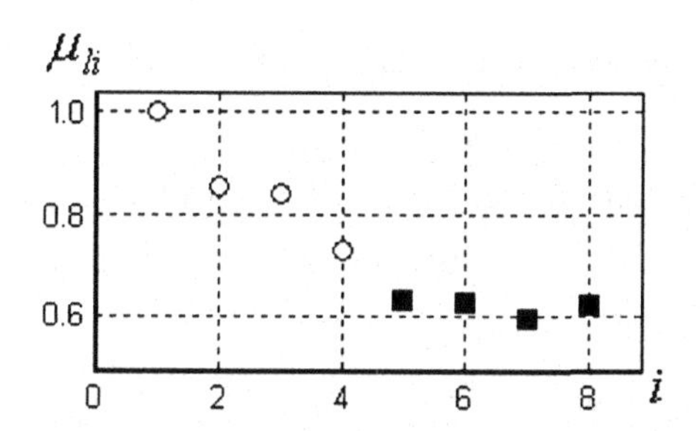

Fig. 3.23 Membership functions of two classes obtained from the D-AFC(c)-algorithm using formulae (3.61) and (1.26)

Therefore, we obtain the following: the first class is formed by four elements and the second class is composed of four elements, too. The first object is the typical point of the first fuzzy cluster and the fifth object is the typical point of the second fuzzy cluster. Note that the second fuzzy cluster $A^5_{(\alpha=0.52647)}$ is a subnormal fuzzy set in the sense of Definition 1.3 because the condition $h(A^5_{(\alpha=0.52647)}) = 0.64035 < 1$ is met. It is should be noted that the membership values of the first class are equal in both experiments.

The detection of most "plausible" fuzzy clusters in the clustering structure sought for the uncertain data set X can be considered as the final aim of classification and the construction of the set of values of the most possible number of fuzzy clusters with their corresponding possibility degrees is an important step in this way. For this purpose, a procedure for the construction of the set of values of the most possible number of fuzzy clusters in some sought structure was proposed in [146]. The procedure is based on the interval-valued data processing by the D-AFC-TC-algorithm of possibilistic clustering by choosing a distance for fuzzy sets. Moreover, the procedure was extended for the case of the three-way data in [148]. Let us assume that the initial data is presented by the poly-matrix (3.50).

There is the following five-step procedure for constructing the set of values of the most possible number of fuzzy clusters:

1. Cosntruct the set of m_2 matrices $X^{t_2}_{n\times m_1} = [x^{t_1}_i]$ after the normalization of the initial data matrix $\hat{X}_{n\times m_1\times m_2} = [\hat{x}^{t_1(t_2)}_i]$ with each matrix processed by the D-AFC-TC-algorithm by choosing a suitable distance $d(x_i, x_j)$; the number $c^{(t_2)}$ of fully separated fuzzy clusters in the allotment $R^*_c(X^{t_2})$ and the corresponding value of tolerance threshold $\alpha^{(t_2)}$ are main results of classification and the results can be described by two sets $\tilde{C} = \{c^{(\hat{t}_2)} \mid \hat{t}_2 = 1,\ldots,\hat{m}_2\}$ and $\tilde{A} = \{\alpha^{(\hat{t}_2)} \mid \hat{t}_2 = 1,\ldots,\hat{m}_2\}$ where $1 \le \hat{m}_2 \le m_2$;

2. Construct the triangular fuzzy number $V^{(\hat{t}_2)} = (m^{(\hat{t}_2)}, a^{(\hat{t}_2)}, b^{(\hat{t}_2)})_T$ for each value $c^{(\hat{t}_2)} \in \tilde{C}$ as follows: $m^{(\hat{t}_2)} = c^{(\hat{t}_2)}$, $a^{(\hat{t}_2)} = m^{(\hat{t}_2)} - 1$, $b^{(\hat{t}_2)} = n - m^{(\hat{t}_2)}$ and membership functions of the fuzzy numbers $V^{(\hat{t}_2)}$, $\hat{t}_2 \in \{1,\ldots,\hat{m}_2\}$ are defined by the expression (1.67) where $x \in (1,n)$ and $\mu_{V^{(\hat{t}_2)}}(1) = \mu_{V^{(\hat{t}_2)}}(n) = 0$;
3. Construct the fuzzy set $\hat{V}^{(\hat{t}_2)} = \{\hat{c}_g, \mu_{\hat{V}^{(\hat{t}_2)}}(\hat{c}_g)\}$ from each triangular fuzzy number $V^{(\hat{t}_2)}$, $\hat{t}_2 \in \{1,\ldots,\hat{m}_2\}$, as follows: a subset of integer values $\hat{C} = \{\hat{c}_*, \ldots, \hat{c}^*\}$, where $\hat{c}_* = 2$ and $\hat{c}^* = n-1$, should be extracted from $(1,n)$ and the value of the membership degree $\mu_{\hat{V}^{(\hat{t}_2)}}(\hat{c}_g)$, $\hat{c}_g \in \hat{C}$, of each fuzzy set $\hat{V}^{(\hat{t}_2)}$ is equal to the value of the membership function $\mu_{V^{(\hat{t}_2)}}(x)$ of the corresponding fuzzy number $V^{(\hat{t}_2)}$ in the case $x = \hat{c}_g$;
4. Construct the fuzzy union $D = \underset{\hat{t}_2}{\mathrm{S}}(\mu_{\hat{V}^{(\hat{t}_2)}}(\hat{c}_g))$ of all fuzzy sets $\hat{V}^{(\hat{t}_2)} = \{\hat{c}_g, \mu_{\hat{V}^{(\hat{t}_2)}}(\hat{c}_g)\}$, $\hat{c}_g \in \hat{C}$, $\hat{t}_2 = 1,\ldots,\hat{m}_2$, with its membership function $\mu_{\underset{\hat{t}_2}{\mathrm{S}}}(\hat{c}_g)$ according to a selected S-norm (1.17) – (1.19);
5. Construct the α-level fuzzy set for the fuzzy set D as follows: $D_{(\alpha)} = \left\{(\hat{c}_g \in D_\alpha, \mu_{D_{(\alpha)}}(\hat{c}_g) = \mu_D(\hat{c}_g))\right\}$, where $D_\alpha = \{\hat{c}_g \in \hat{C} \mid \mu_D(\hat{c}_g) \geq \hat{\alpha}\}$ and $\hat{\alpha} = \min\limits_{\hat{t}_2} \alpha^{(\hat{t}_2)}$.

The set $D_\alpha = Supp(D_{(\alpha)})$ is the set of values of the most possible number of fuzzy clusters in some clustering structure sought. So, the bounds $c_{\min}$ and $c_{\max}$ for the number of clusters c can be estimated. The membership function $\mu_{D_{(\alpha)}}(\hat{c}_g)$ can be interpreted as a possibility distribution π [146] and the possibility degrees $\pi(\hat{c}_g)$ express the extent to which the number $\hat{c}_g \in D_\alpha$ of fuzzy clusters is plausible. It should be noted that the presented procedure can be generalized for the case of LR-fuzzy numbers $V^{(\hat{t}_2)} = (m^{(\hat{t}_2)}, a^{(\hat{t}_2)}, b^{(\hat{t}_2)})_{LR}$ in the sense of the formula (1.65).

A technique of constructing the stationary clustering structure for the uncertain data set can be devised as a two-step process in which the set D_α of values of the most possible number of fuzzy clusters is a preliminary result of classification. The allotment $R_c^*(X)$ among an a priori unknown number of fuzzy clusters can be considered as the clustering structure sought.

The technique of for the construction of the allotment among fuzzy clusters for the uncertain data set proposed in [148] can be outlined as follows:

1. The initial data are contained in the poly-matrix of attributes $\hat{X}_{n \times m_1 \times m_2} = [\hat{x}_i^{t_1(t_2)}]$, $i = 1,\ldots,n$, $t_1 = 1,\ldots,m_1$, $t_2 = 1,\ldots,m_2$, and the procedure of constructing the set D_α of values of the most possible number of fuzzy clustersis applied to this data set;
2. The matrix of tolerance coefficients $T = [\mu_T(x_i, x_j)]$, $i, j = 1,\ldots,n$, is constructed from the normalized initial data by choosing a suitable distance for the type-two or interval-valued fuzzy sets;
3. The D-AFC(c)-algorithm, using some cluster validity index, is applied directly to the matrix of tolerance coefficients for the set D_α, and the allotment $R_c^*(X)$ is the final result of the classification process.

The allocation of objects among the a priori unknown number of fuzzy clusters, which is the result of application of the proposed technique to the initial uncertain data set, is appropriate for any current values of the measured quantities $\hat{x}_i^{t_1(t_2)}$. It belongs either to the interval $[\hat{x}_i^{t_1(\min)}, \hat{x}_i^{t_1(\max)}]$, $t_2 \in \{\min, \max\}$, in the situation of interval uncertainty, or to the set $(\hat{x}_i^{t_1(1)}, \ldots, \hat{x}_i^{t_1(t_2)}, \ldots, \hat{x}_i^{t_1(m_2)})$, $t_2 \in \{1,\ldots,m_2\}$, in the case of the three-way data for all objects $x_i \in X$, $t_1 = 1,\ldots,m_1$. So, the obtained allotment $R_c^*(X)$ among the discovered number c of fuzzy clusters can be considered as the stationary clustering structure.

3.4.3 An Illustrative Example

An application of the proposed technique to classification problem can be illustrated by Ichino and Yaguchi's [56] oil data set.

The original data set consists of the specific values of gravity, freezing point, iodine value, saponification value, and major fatty acids measured for 8 types of oils. It should be noted that the specific gravity, freezing point, iodine value, and saponification are quantitative attributes which take on interval values and the fatty acids are the nominal attributes which takes finite sets as their values. However, the specific gravity, iodine value, and saponification value were selected as the object attributes for the experiment. The data set is presented in Table 3.11.

Table 3.11 Ichino and Yaguchi's [56] oil data set

Object number , i	Types of oils	Specific gravity	Iodine value	Saponification value
1	Linseed oil	0.930-0.935	170-204	118-196
2	Perilla oil	0.930-0.935	192-208	188-197
3	Cottonseed oil	0.916-0.918	99-113	189-198
4	Sesame oil	0.920-0.926	104-116	187-193
5	Camellia oil	0.916-0.917	80-82	189-193
6	Olive oil	0.914-0.919	79-90	187-196
7	Beef tallow	0.860-0.870	40-48	190-199
8	Hog fat	0.858-0.864	53-77	190-202

The analysis of types of oils shown in Table 3.11 highlights that the first six oils are vegetable oils and the remaining two are the animal oils. That is why we expect to find two clusters in this data set.

The data matrix $\hat{X}_{8\times3\times2} = [\hat{x}_i^{t_1(t_2)}]$, $i = 1,\ldots,8$, $t_1 = 1,\ldots,3$, $t_2 \in \{\min,\max\}$, was normalized according to the formula (3.52). The procedure for constructing the set of values of the most possible number of fuzzy clusters was applied to the normalized data using the squared normalized Euclidean distance (2.15) for the D-AFC-TC-algorithm and the maximum operation (1.17) as the fuzzy union. By executing the D-AFC-TC-algorithm for the matrix of the minimum values $X_{8\times3}^{\min} = [x_i^{t_1}]$, we obtain that the allotment $R_c^*(X^{\min})$ is formed by two fuzzy clusters and it was obtained for the tolerance threshold $\alpha^{(\min)} = 0.7899$. On the other hand, by executing the D-AFC-TC-algorithm for the matrix of the maximum values $X_{8\times3}^{\max} = [x_i^{t_1}]$ we obtain that the allotment $R_c^*(X^{\max})$ includes three fuzzy clusters and it was obtained for the tolerance threshold $\alpha^{(\max)} = 0.9876$. That is why the clustering structure of Ichino and Yaguchi's oil data set is an unstable clustering structure. The set of values of the most possible number of fuzzy clusters in the sought allotment $R_c^*(X)$ is $D_\alpha = \{c_{\min} = 2,\ldots,c_{\max} = 4\}$ and corresponding possibility degrees are shown in Figure 3.24.

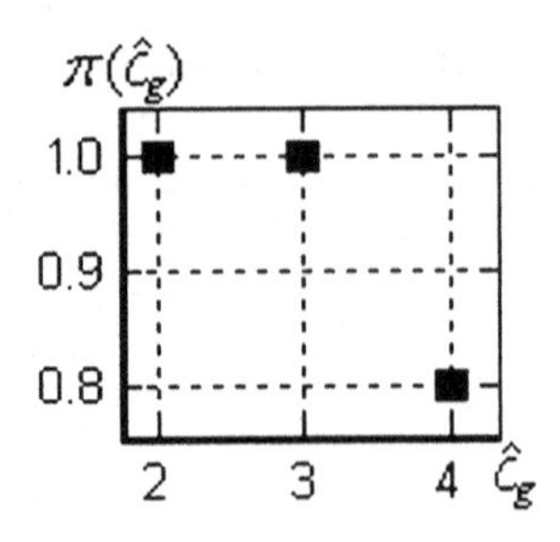

Fig. 3.24 Possibility degrees obtained using the maximum operation (1.17)

The set D_α and the possibility degrees $\pi(c_g)$ depend on the type of the fuzzy union. If the probabilistic sum (1.18) or the bounded sum (1.19) are selected as the fuzzy union, then the set of values of the most possible number of fuzzy clusters is $D_\alpha = \{c_{\min} = 2, \ldots, c_{\max} = 5\}$. The possibility degrees obtained by using the probabilistic sum (1.18) are presented in Figure 3.25.

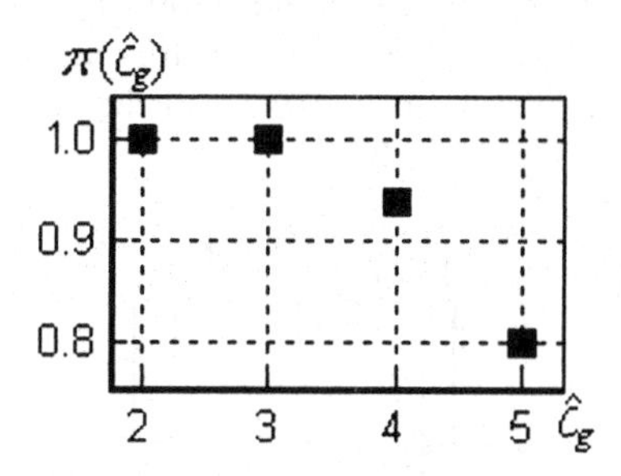

Fig. 3.25 Possibility degrees obtained by using the probabilistic sum (1.18)

The dissimilarity function (3.62) was used for data preprocessing in the first experiment. The matrix of the corresponding feeble fuzzy intolerance relation I_1 is presented in Table 3.12 where each object of the data set is denoted by x_i, $i \in \{1, \ldots, 8\}$.

Table 3.12 Matrix of dissimilarity coefficients obtained using the dissimilarity function (3.62)

I_1	x_1	x_2	x_3	x_4	x_5	x_6	x_7	x_8
x_1	0.15123							
x_2	0.13921	0.00412						
x_3	0.23114	0.12156	0.00318					
x_4	0.20351	0.10445	0.00573	0.00271				
x_5	0.27753	0.18402	0.00949	0.01392	0.00042			
x_6	0.27198	0.17572	0.00891	0.01318	0.00203	0.00333		
x_7	0.64226	0.54837	0.20160	0.24543	0.16862	0.17312	0.00510	
x_8	0.61457	0.50920	0.20195	0.24699	0.18150	0.18419	0.01267	0.00781

It is should be noted that the condition of weak antireflexivity (1.40) is not met for the matrix of feeble fuzzy intolerance $I_1 = [\mu_{I_1}(x_i, x_j)]$, $i, j = 1, \ldots, 8$. For example, the condition $\mu_{I_1}(x_1, x_2) < \mu_{I_1}(x_1, x_1)$ is met for the objects x_1 and x_2. Moreover, the condition $\mu_{I_1}(x_i, x_i) > 0$ is met for all objects x_i, $i = 1, \ldots, 8$.

The complement operation (1.26) was applied to the matrix of fuzzy intolerance $I_1 = [\mu_{I_1}(x_i, x_j)]$, $i, j = 1,\ldots,8$. So, the matrix $T_1 = [\mu_{T_1}(x_i, x_j)]$, $i, j = 1,\ldots,8$ of feeble fuzzy tolerance was obtained and the corresponding feeble fuzzy tolerance is a subnormal fuzzy relation.

By executing the D-AFC(c)-algorithm for $\hat{c}_g \in D_\alpha = \{c_{\min} = 2,\ldots,c_{\max} = 4\}$ using the quadratic measure of fuzziness of the allotment (2.31), we obtain that the current number of fuzzy clusters is equal 2 and this number corresponds to the maximum of the quadratic measure of fuzziness of the allotment (2.31). The plot of this measure is shown in Figure 3.26.

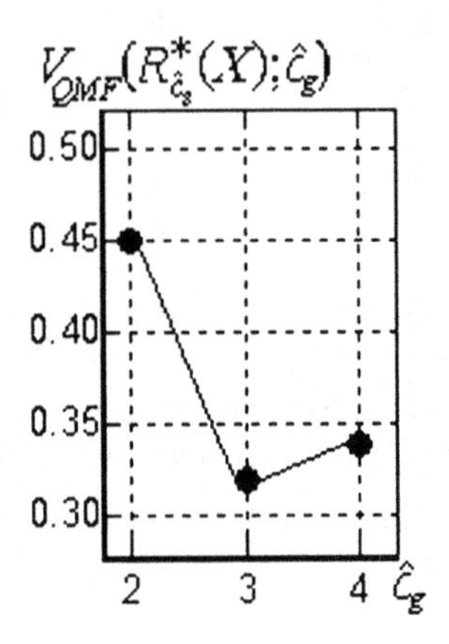

Fig. 3.26 Plot of the quadratic measure of fuzziness for Ichino and Yaguchi's oil data set using the function of dissimilarity (3.62)

The membership functions of two classes of the allotment $R_c^*(X)$ are presented in Figure 3.27 in which the membership values of the first class are represented by ○ and the membership values of the second class are represented by ■.

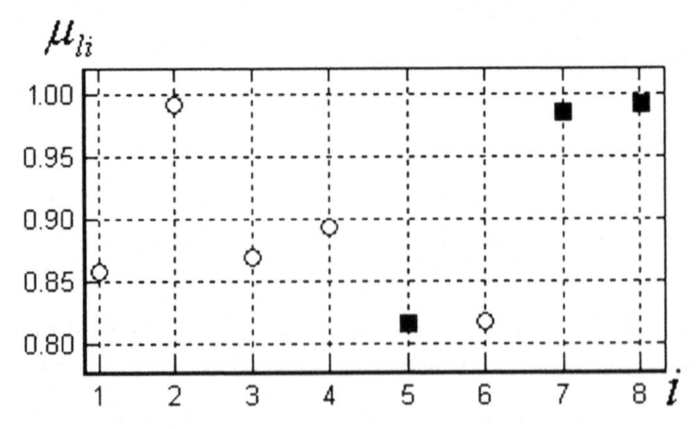

Fig. 3.27 Membership functions of two classes obtained from the D-AFC(c)-algorithm using the function of dissimilarity (3.62)

The allotment $R_c^*(X)$ among two fully separated fuzzy clusters was obtained for the tolerance threshold $\alpha = 0.8184$. The first class is formed by 5 elements and the second class by 3 elements. So, the fifth element is a misclassified object. The second object is the typical point of the first fuzzy cluster and the eighth object is the typical point of the fuzzy cluster which corresponds to the second class. It should be noted that both fuzzy clusters are subnormal fuzzy sets.

By executing the D-AFC(c)-algorithm for $c_g \in D_\alpha = \{c_{\min} = 2, \ldots, c_{\max} = 4\}$ using Ju and Yuan's [57] similarity measure (3.58) for $\lambda = 2$ and the validity measure (2.31), we obtain that the numbers $c = 3$ and $c = 4$ of fuzzy clusters in the allotment sought $R_c^*(X)$ are suboptimal. Thus, the allotment $R_c^*(X)$ among two fully separated fuzzy clusters was obtained for the tolerance threshold $\alpha = 0.5852$. The first class is formed by 6 elements and the second class by 2 elements. So, the misclassified objects are absent at the resulting allotment $R_c^*(X)$.

The membership functions of two classes of the allotment $R_c^*(X)$ are presented in Figure 3.28. The membership values of the first class are represented by ○ and membership values of the second class are represented by ■.

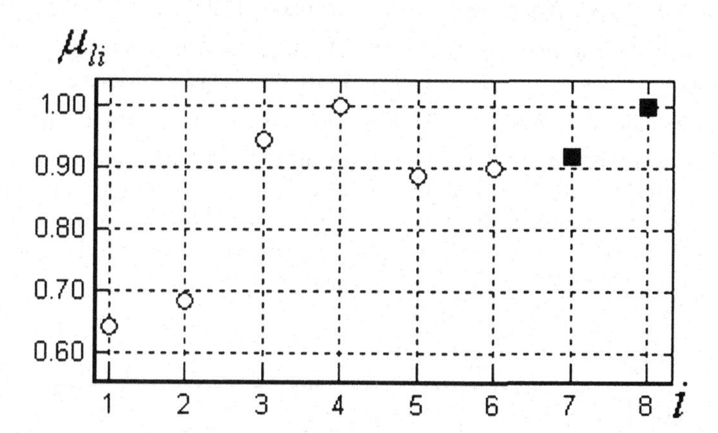

Fig. 3.28 Membership functions of two classes obtained from the D-AFC(c)-algorithm using the similarity measure (3.58)

So, both fuzzy clusters are the normal fuzzy sets. The value of the membership function of the fuzzy cluster which corresponds to the first class is maximal for the fourth object and is equal one. So, the fourth object is the typical point of the first fuzzy cluster. The membership value of the eighth object is equal one for the second fuzzy cluster. Thus, the eighth object is the typical point of the second fuzzy cluster.

By executing the D-AFC(c)-algorithm using the distance between the interval-valued fuzzy sets (3.59) and the validity measure (2.31), we obtain that the number $c = 4$ of fuzzy clusters in the allotment sought $R_c^*(X)$ is suboptimal and

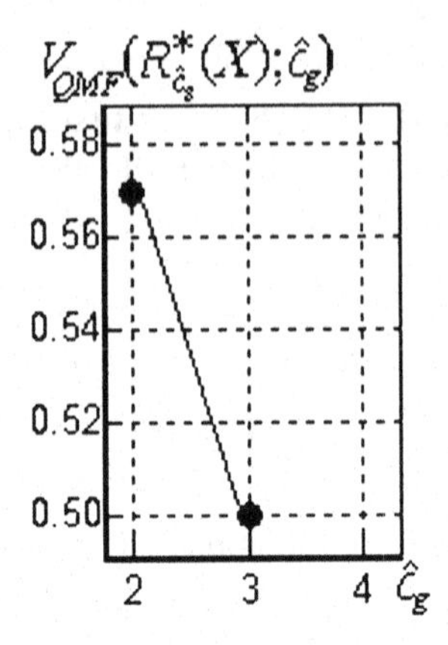

Fig. 3.29 Plot of the quadratic measure of fuzziness for Ichino and Yaguchi's oil data set using the distance (3.59)

the actual number of fuzzy clusters is equal 2. The performance of this validity measure is shown in Figure 3.29.

The allotment $R^*_c(X)$ among two fully separated fuzzy clusters was obtained for the tolerance threshold $\alpha = 0.5243$. The first class is formed by 1 element and the second class includes 7 elements. So, five misclassifications are present in the resulting allotment $R^*_c(X)$. The first object is the typical point of the first fuzzy cluster and the sixth object is the typical point of the second fuzzy cluster. Both fuzzy clusters are the normal fuzzy sets.

The membership functions of two classes of the allotment are presented in Figure 3.30 where the membership values of the first class are represented by ○ and the membership values of the second class are represented by ■.

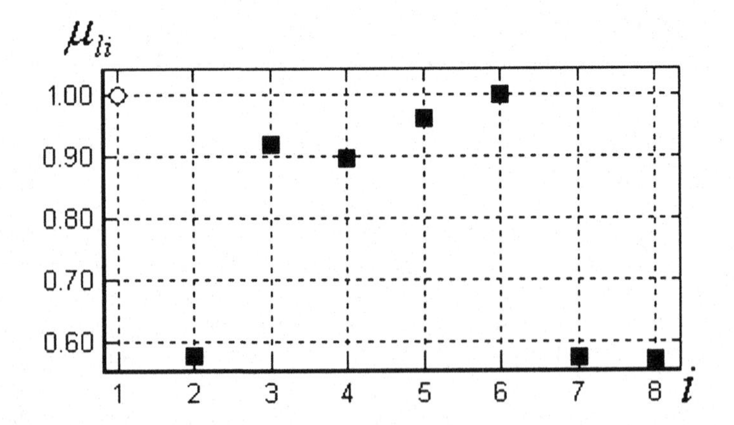

Fig. 3.30 Membership functions of two classes obtained from the D-AFC(c)-algorithm using the distance (3.59)

The results obtained by using this technique depend on the selection of a dissimilarity measure and an initial data normalization method. Moreover, the set of values of the most possible number of fuzzy clustersin the clustering structure sought depends on the type of the selected S-norm. So, we can conclude that the use of some dissimilarity measure may imply some objections. It will therefore be reasonable to make use of various dissimilarity measures and compare the obtained clustering results.

The proposed technique of constructing the stationary clustering structure can be very simply generalized to the case of the fuzzy c -partition (1.71) [147]. For this purpose, the matrix of dissimilarity coefficients $I=[\mu_I(x_i,x_j)]$, $i,j=1,\ldots,n$, was constructed from the normalized initial data in the second step of the described technique of constructing the stable clustering structure for the uncertain data set and the relational ARCA-algorithm [25] of fuzzy clustering, using the partition coefficient (1.89), was applied directly to the obtained matrix in the third step.

Let us consider the results of experiments which are presented in [147]. In all experiments we assume that the number of fuzzy clusters is $\hat{c}_g \in D_\alpha = \{c_{\min}=2,\ldots,c_{\max}=4\}$ and the value of the weighting exponent is $\gamma=2$.

The matrix of dissimilarity coefficients constructed according to the function of dissimilarity (3.62) was used in the first experiment. The plot of the partition coefficient is presented in Figure 3.31.

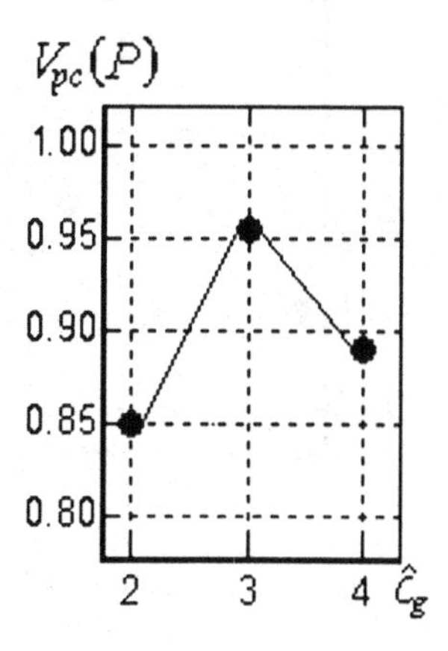

Fig. 3.31 Plot of the partition coefficient for Ichino and Yaguchi's oil data set using the distance (3.62)

Ju and Yuan's similarity measure (3.58) for $\lambda=2$ was used in the second experiment. Thus, the matrix of fuzzy intolerance $I=[\mu_I(x_i,x_j)]$ was obtained after the application of the complement operation (1.26) to the matrix of fuzzy intolerance $T=[\mu_T(x_i,x_j)]$. The plot of the validity measure (1.89) is shown in Figure 3.32.

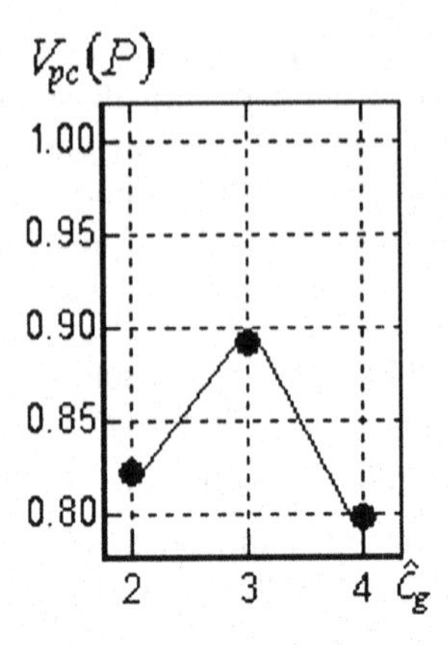

Fig. 3.32 Plot of the partition coefficient for Ichino and Yaguchi's oil data set using the similarity measure (3.58) and the complement operation (1.26)

Grzegorzewski's [43] distance between the interval-valued fuzzy sets (3.59) was used in the third experiment. The plot of the partition coefficient is presented in Figure 3.32.

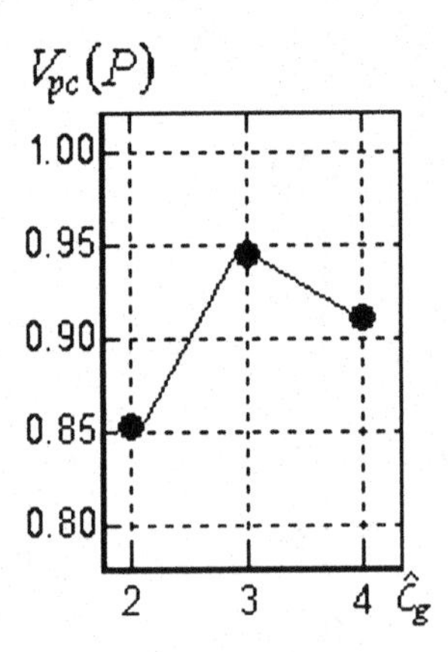

Fig. 3.33 Plot of the partition coefficient for Ichino and Yaguchi's oil data set using the distance (3.59)

So, the number of fuzzy clusters is equal 3 in all experiments and this number corresponds to the maximum partition coefficient (1.89). The matrix $P_{3\times 8} = [u_{li}]$ of the fuzzy c-partition $P(X)$ and the matrix of dissimilarities $K_{c\times n} = [d_{li}]$ between the prototypes τ^l, $l = 1,\ldots,3$ and objects x_i, $i = 1,\ldots,8$, are the results of each experiment.

The ARCA-algorithm using the the partition coefficient (1.89) can therefore determine the number of fuzzy clusters which seems to be more suitable for some specific applications.

Moreover, the membership values u_{li} and the values of the dissimilarity coefficients $d(x_i, \tau^l)$ between the prototypes τ^l, $l = 1,\ldots,3$ and the objects x_i, $i = 1,\ldots,8$ are very similar in each experiment. For example, Figure 3.34 shows the membership functions of three classes of the fuzzy c-partition $P(X)$ which is obtained from the ARCA-algorithm in the first experiment. The membership values of the first class are represented by ○, the membership values of the second class are represented by ■, and membership values of the third class are represented by ∇.

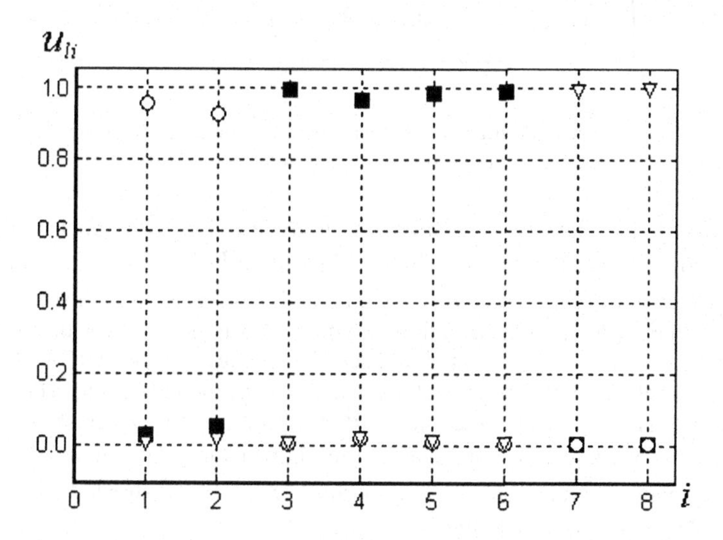

Fig. 3.34 Membership functions of three fuzzy clusters obtained from the ARCA-algorithm using the function of dissimilarity (3.62)

The values of the dissimilarity coefficients $d(x_i, \tau^l)$, $i = 1,\ldots,8$, for three classes $A^l \in P(X)$, $l = 1,\ldots,3$, are shown in Figure 3.35. The values of the dissimilarity coefficients for the first class objects are represented by ○, those for the second class objects are represented by ■, and the values of dissimilarities for the second class objects are represented by ∇.

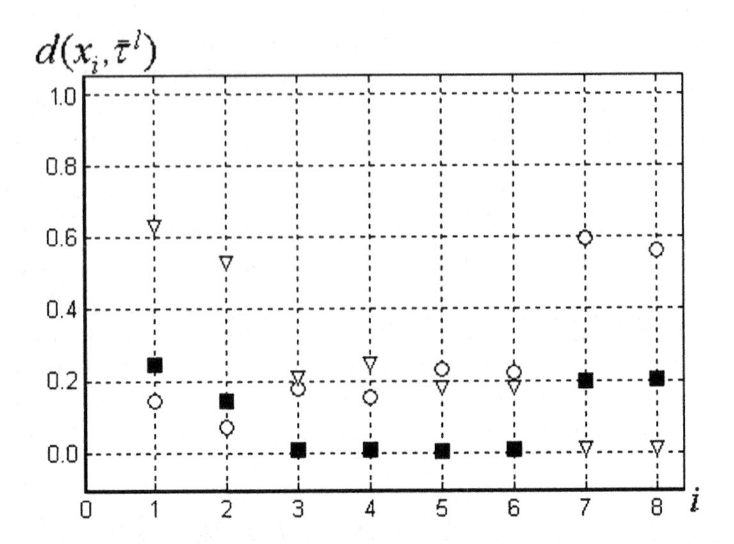

Fig. 3.35 Values of dissimilarities between prototypes of three fuzzy clusters and objects obtained from the ARCA-algorithm using the function of dissimilarity (3.62)

Obviously, the class of vegetables oils is divided into two subclasses in the considered case. So, the obtained fuzzy c-partition $P(X)$ can be considered as the stationary clustering structure for Ichino and Yaguchi's oil data set. It should be noted that the condition of antireflexivity (1.38) is not met for the matrix of dissimilarity coefficients presented in Table 3.12. However, the ARCA-algorithm has been applied to the relational data and the presented result seems to be appropriate. That is why the relational ARCA-algorithm of fuzzy clustering can be applied for constructing the stationary clustering structure $P(X)$ in the case of the three-way data. The functions of dissimilarity (3.53) – (3.55) should be used in the case for the three-way data preprocessing.

In general, the results of application of the proposed technique of constructing the stationary clustering structure to the oil data set show that the technique is an effective and efficient tool for solving the classification problem under uncertainty of the initial data.

Chapter 4
Applications of Heuristic Algorithms of Possibilistic Clustering

Clustering methods can be an appropriate tool for solving different problems related to classification, such as the construction of the set of labeled objects for semi-supervised fuzzy clustering algorithms, feature selection, classification of the asymmetric data, fuzzy classifier prototyping, decision making, etc. The proposed approach to possibilistic clustering shows a close relationship to some of the problems mentioned above.

In this chapter we will outline some techniques that are needed and useful for the application of the proposed clustering approach to solving such problems.

4.1 A Methodology of Fuzzy Clustering with Partial Supervision

The present section describes a two-step technique of partially-supervised fuzzy clustering. Pedrycz's [89] algorithm of fuzzy clustering and the D-AFC(c)-algorithm form the basis of this technique. In the first subsection some basic concepts of Pedrycz's clustering algorithm and a methodology of fuzzy clustering with partial supervision are considered. The second subsection includes an example of the application of the method considered to Anderson's Iris data.

4.1.1 Constructing the Set of Labeled Objects

A priori knowledge about the belongingness of some objects can often be very useful for classification. This fact has been a basis of an approach to fuzzy clustering with partial supervision. Algorithms of fuzzy clustering with partial supervision have been proposed by Pedrycz [88]. Numerical experiments have shown that knowledge concerning the membership of a small portion of the patterns significantly improves the results of clustering in such a sense that the partition matrix detects a real structure existing in the data set. Moreover, the speed of convergence of the scheme has been improved. These facts have been demonstrated by Pedrycz [89].

The basic concept of Pedrycz 's [88] algorithm of fuzzy clustering in the presence of labeled patterns should first be considered. The algorithm has been

D.A. Viattchenin: *A Heuristic Approach to Possibilistic Clustering*, Studfuzz 297, pp. 183–218.
DOI: 10.1007/978-3-642-35536-3_4

called the ssfcm-algorithm by Bezdek, Keller, Krishnapuram and Pal [11]. Let us consider a subset of labeled objects $X_L \subset X$. The elements of this subset will be indicated by the Boolean vector $s = (s_1, s_2, \ldots, s_n)^T$ with the entries

$$s_i = \begin{cases} 1, if \ x_i \in X_L \\ 0, if \ x_i \notin X_L \end{cases}. \tag{4.1}$$

Let Y be the partition matrix $Y_{c \times n} = [y_{li}]$, $l = 1, \ldots, c$, $i = 1, \ldots, n$, in which the elements of Y are assigned as follows: if $x_i \in X_L$, then the values of y_{li} are given by the analyst and, additionally, the condition $\sum_{l=1}^{c} y_{li} = 1$ must be met. Otherwise, the values of y_{li} may be normalized in order to fulfill this condition. If $x_i \notin X_L$, then the values of the respective column of Y are irrelevant. So, the additional information is conveyed by the labeled pattern. Then the performance index (1.72) has been modified by Pedrycz [88] as follows:

$$Q_{ssfcm(1)}(P, \mathrm{T}) = \sum_{l=1}^{c} \sum_{i=1}^{n} u_{li}^2 d^2(x_i, \tau^l) + \sum_{l=1}^{c} \sum_{i=1}^{n} (u_{li} - y_{li} s_i)^2 d^2(x_i, \tau^l), \tag{4.2}$$

where u_{li}, $l = 1, \ldots, c$, $i = 1, \ldots, n$, is the membership degree, x_i, $i \in \{1, \ldots, n\}$, is the data point, $\mathrm{T} = \{\tau^1, \ldots, \tau^c\}$ is the set fuzzy cluster prototypes, $d^2(x_i, \tau^l)$ is the squared Euclidean distance (1.73), the value of the weighting exponent is equal two, $\gamma = 2$.

In comparison to (1.72) the objective function (4.2) takes into account the weighted sum of differences between the grades of membership assigned to these objects by the clustering algorithm u_{li} and the grades of membership of the labeled patterns y_{li}.

Modifications of the criterion (4.2) were also proposed by Pedrycz [88]. In the first place, the criterion (4.2) was modified in such a way so as to attain a balanced influence from the labeled and unlabeled patterns on the computed partition matrix:

$$Q_{ssfcm(2)}(P, \mathrm{T}) = a_1 \sum_{l=1}^{c} \sum_{i=1}^{n} u_{li}^2 d^2(x_i, \tau^l) + a_2 \sum_{l=1}^{c} \sum_{i=1}^{n} (u_{li} - y_{li} s_i)^2 d^2(x_i, \tau^l), \tag{4.3}$$

where $a_1 = 1/(n - n_L)$, $a_2 = 1/n_L$, n_L is the number of labeled patterns, and $d^2(x_i, \tau^l)$ is the squared Euclidean distance (1.73). The second modification of the criterion (4.2) is the objective function (4.3), where $d^2(x_i, \tau^l)$ is the squared Mahalanobis distance. The partially supervised approach to fuzzy clustering based on the criteria (4.2) and (4.3) was generalized by Pedrycz and Waletzky [91].

In [89] several numerical examples dealing with real and simulated data have been studied. These examples show that the partial supervision is very effective and efficient in real applications of fuzzy clustering. However, a fuzzy c-partition is sensitive to the choice of the labeled objects. This fact was demonstrated in [113]. So, the problem of construction of the set of labeled objects $X_L \subset X$ is very important for a fruitful application of Pedrycz's fuzzy clustering method to a real classification problem solving. The results of application of heuristic algorithms of possibilistic clustering can be very well interpreted and the clustering results are stable. This fact is very relevant for the construction of the subset of labeled objects. The typical points of fuzzy clusters which are elements of the allotment can be selected as the labeled objects for the application of Pedrycz's algorithm to the data in the classification process. The method of fuzzy clustering with partial supervision was proposed in [124] and the methodology can be described as follows:

1. The initial data are contained in the matrix of attributes $\hat{X}_{n \times m_1} = [\hat{x}_i^{t_1}]$, $i = 1, \ldots, n$, $t_1 = 1, \ldots, m_1$. The matrix of tolerance coefficients $T = [\mu_T(x_i, x_j)]$, $i, j = 1, \ldots, n$, can be constructed by normalizing the initial data and choosing a suitable distance for the fuzzy sets;
2. The D-AFC(c)-algorithm can be applied directly to the matrix of tolerance coefficients and an allotment $R_c^*(X) = \{A_{(\alpha)}^1, \ldots, A_{(\alpha)}^c\}$ is the result of the first step of the classification process;
3. The typical points τ_e^l, $l = 1, \ldots, c$, of the fuzzy clusters $A_{(\alpha)}^l$, $l = 1, \ldots, c$ of the obtained allotment $R_c^*(X)$ can be selected as the labeled objects and the elements of the partition matrix $Y_{c \times n} = [y_{li}]$, $l = 1, \ldots, c$, $i = 1, \ldots, n$, can be assigned as follows: if $x_i \in X_L$ and $x_i = \tau_e^l$, then the value of y_{li} is equal one; otherwise, if $x_i \notin X_L$ the value y_{li} is equal zero;
4. Pedrycz's fuzzy clustering ssfcm-algorithm is applied directly to the matrix of attributes $\hat{X}_{n \times m_1} = [\hat{x}_i^{t_1}]$, $i = 1, \ldots, n$, $t_1 = 1, \ldots, m_1$, for the set $X_L = \{\tau_e^l \mid l = 1, \ldots, c\}$.

So, the approach of fuzzy clustering with partial supervision can be considered as a two-step fuzzy clustering process in whch the allotment $R_c^*(X)$ is a preliminary result of classification and the fuzzy c-partition $P(X)$ is the final result of classification.

4.1.2 An Illustrative Example

An application of the described technique to a classification problem can be illustrated on Anderson's Iris data. The clustering results, obtained from the D-AFC(c)-algorithm, were considered above. Obviously, the typical points of fuzzy clusters in the allotment $R_c^*(X)$ depend on the distance used in data preprocessing. That is why three experiments were made. Formula (2.11) was used for data normalization in each experiment. Let us consider the results of these experiments.

By executing the D-AFC(c)-algorithm for three classes using the normalized Hamming distance (2.13) in the process of data preprocessing, we obtain that the objects x_{95}, x_{98} and x_{126} are the typical points τ^l, $l = 1,\ldots,3$, of the fuzzy clusters $A_{(\alpha)}^l \in R_{c=3}^*(X)$, $l = 1,\ldots,3$. These objects constitute the subset of labeled objects $X_L \subset X$ in the first experiment and the membership function values y_{li}, $l = 1,\ldots,3$, are equal one for each labeled object.

By executing the D-AFC(c)-algorithm for three classes using the normalized Euclidean distance (2.14) in the process of data preprocessing, we obtain that the objects x_{95}, x_{98} and x_{23} constitue the subset of labeled objects $X_L \subset X$ in the second experiment. The values of the membership function y_{li}, $l = 1,\ldots,3$, are equal one for all $x_i \in X_L$.

By executing the D-AFC(c)-algorithm for three classes using the squared normalized Euclidean distance (2.15) in the process of data preprocessing, we construct the subset of labeled objects $X_L = \{x_{95}, x_{98}, x_{73}\}$ in the third experiment and the membership function values y_{li}, $l = 1,\ldots,3$, are equal one for each element of the subset of labeled objects.

The fuzzy c-partition $P(X)$ was obtained after the application of Pedrycz's algorithm to Anderson's Iris data for the corresponding subset of labeled objects in each experiment. The grades of membership u_{li}, $l = 1,\ldots,3$, $i = 1,\ldots,150$, are in general similar in all three experiments. For example, Figure 4.1 shows the membership function values u_{li}, $i = 1,\ldots,150$, of three classes A^l, $l = 1,\ldots,3$, of the fuzzy c-partition $P(X)$ which was obtained by the application of Pedrycz's algorithm to Anderson's Iris data for the subset of labeled objects $X_L = \{x_{95}, x_{98}, x_{73}\}$.

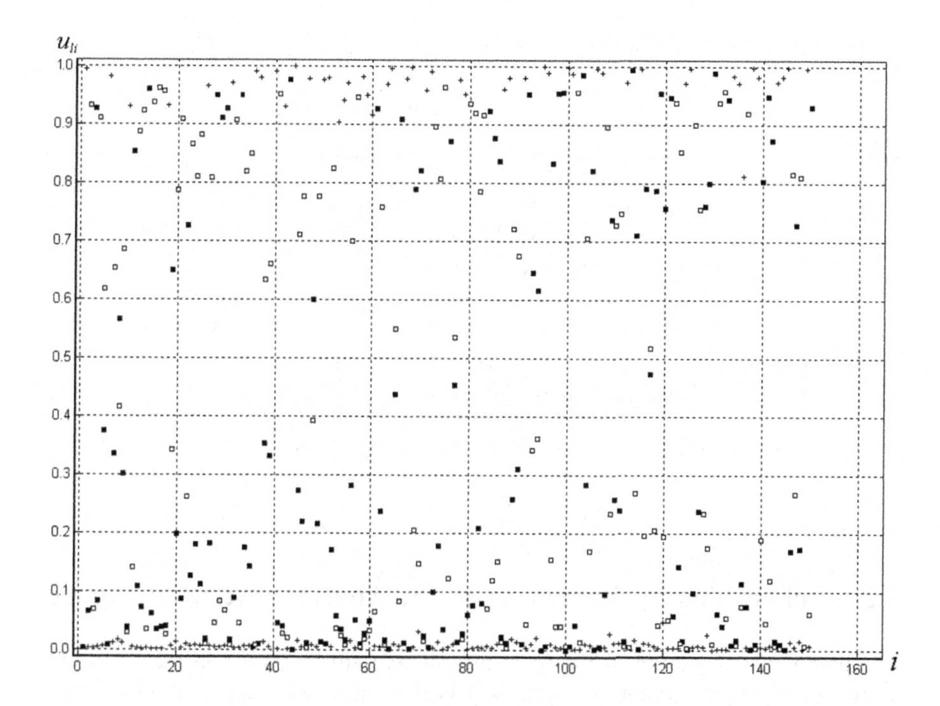

Fig. 4.1 Membership values of the fuzzy partition obtained for the labeled objects x_{95}, x_{98}, x_{73}

The first class corresponds to Iris Setosa. The second class corresponds to Iris Versicolor and the third class corresponds to Iris Virginica. The membership values of the first class are represented by +, the membership values of the second class are represented by ■, and the membership values of the third class are represented by □.

If the maximum memberships rule (1.96) is applied to the matrix $P_{3\times150} = [u_{li}]$, $l = 1,\ldots,3$, $i = 1,\ldots,150$, of the fuzzy c-partition $P(X)$, then the application of Pedrycz's algorithm to Anderson's Iris data yields similar results in each experiment described before. The object assignments resulting from the use of the described method to Anderson's Iris data are presented in Table 4.1.

The misclassified objects are distinguished in Table 4.1. So, there are five mistakes of classification only in each experiment. That is why the results obtained can be considered to be precise and stable.

To briefly summarize, the results of application of the method to Anderson's Iris data show that the methodology offers a precise, and effective and efficient procedure for solving the classification problem. Obviously, the proposed approach to data clustering can be very useful in detailed data analyses too.

Table 4.1 Results of using the methodproposed to Iris data: object assignment

Class		Numbers of objects
Number	Name	
1	SETOSA	1, 6, 10, 18, 26, 31, 36, 37, 40, 42, 44, 47, 50, 51, 53, 54, 55, 58, 59, 60, 63, 64, 67, 68, 71, 72, 78, 79, 87, 88, 91, 95, 96, 100, 101, 106, 107, 112, 115, 124, 125, 134, 135, 136, 138, 139, 143, 144, 145, 149
2	VERSICOLOR	3, 8, 11, 14, 19, 22, 28, 29, 30, 33, 43, 48, 61, 66, 69, 70, 76, 84, 85, 86, 92, 93, 94, 97, 98, 99, 103, 105, 109, 113, 114, 116, 118, 119, 120, 121, 128, 129, 130, 133, 140, 141, 142, 147, 150
3	VIRGINICA	2, 4, 5, 7, **9**, **12**, 13, 15, 16, 17, 20, 21, 23, 24, 25, 27, 32, 34, 35, **38**, 39, 41, 45, 46, 49, 52, 56, 57, 62, **65**, 73, 74, 75, 77, 80, 81, 82, 83, 89, 90, 102, 104, 108, 110, 111, **117**, 122, 123, 126, 127, 131, 132, 137, 146, 148

4.2 Dimensionality Reduction of the Analyzed Attribute Space

The problem of selection of the most informative attributes is considered in this section. In the first subsection a method based on the application of the D-PAFC-algorithm to solving the problem of dimensionality reduction of the analyzed attributes space is described. The methodology is illustrated on Anderson's Iris data example with a comparison with the result obtained by using the conventional principal component analysis (PCA) will be presented in the second subsection.

4.2.1 A Method of Attribute Clustering

Problems of data visualization and the reduction of dimensionality of the analyzed space of attributes are very important in the process of data analysis. The attribute selection is referred here to the problem of dimensionality reduction of the data which initially contain a high number of attributes. The selection aims at the choice of the minimal number of the original attributes which still contain information that is essential for discovering substructures in the data, while reducing the computational complexity imposed by using many attributes. Different attribute selection methods were briefly described by Ghazavi and Liao [41].

Clustering methods can be applied to solving the problem of attribute selection. In particular, by combining the attributes selection with attribute weights and the semi-supervised fuzzy clustering has been proposed by Kong and Wang [64] in machine learning. On the other hand, the attribute clustering method has been proposed by Au, Chan, Wong and Wang [6]. The corresponding ACA-algorithm finds c disjoint clusters and assigns each attribute to one of the clusters. The attributes in each cluster should have a high correlation with each other while a low correlation with attributes in other clusters. This method uses an interdependence redundancy measure as the similarity measure.

A fuzzy attribute clustering method proposed by Chitsaz, Taheri, and Katebi [20] and the method mentioned above combines the effectiveness and efficiency of the ACA-algorithm with those of the FCM-algorithm. In the FACA-algorithm, each attribute is assigned to different fuzzy clusters with different grades of membership. This comes from the basic idea that each attribute cannot belong to just one cluster and it is much better to consider the correlation of each attribute to other attributes in each cluster. So, accurate relations between the attributes are available during the selection of the most relevant attributes.

An extension of the FACA-algorithm has been considered by Chitsaz, Taheri, Katebi, and Jahromi [21] where four different techniques for the step of attribute selection are introduced. By applying the chi-square test, this approach considers the dependence of each attribute on class labels in the process of attribute selection.

The D-PAFC-algorithm can be applied to solve the problem of attribute clustering. The basic idea of the approach is that attributes can be classified and a typical point of each fuzzy cluster can be considered as an informative attribute. So, a method for solving the problem of reduction of attribute space dimensionality has been proposed in [135] and it can be described as the following three-step procedure:

1. The initial data are contained in the matrix of attributes $\hat{X}_{n \times m_1} = [\hat{x}_i^{t_1}]$, $i = 1, \ldots, n$, $t_1 = 1, \ldots, m_1$. So, the matrix of correlation coefficients $r_{m_1 \times m_1} = [r(\hat{x}^{t_1}, \hat{x}^{k_1})]$, $t_1, k_1 = 1, \ldots, m_1$, is constructed as follows:

$$r(\hat{x}^{t_1}, \hat{x}^{k_1}) = \frac{c_{t_1 k_1}}{s^{t_1} s^{k_1}}, \tag{4.4}$$

where $c_{t_1 k_1} = \frac{1}{n} \sum_{i=1}^{n} \left(\hat{x}_i^{t_1} - \bar{\hat{x}}^{t_1}\right)\left(\hat{x}_i^{k_1} - \bar{\hat{x}}^{k_1}\right)$, $(s^{t_1})^2 = \frac{1}{n} \sum_{i=1}^{n} \left(\hat{x}_i^{t_1} - \bar{\hat{x}}^{t_1}\right)^2$, and $\bar{\hat{x}}^{t_1} = \frac{1}{n} \sum_{i=1}^{n} \hat{x}_i^{t_1}$;

2. The matrix of correlation coefficients for the data is normalized as follows:

$$\tilde{r}(\hat{x}^{t_1}, \hat{x}^{k_1}) = \frac{\left(r(\hat{x}^{t_1}, \hat{x}^{k_1}) - \min\limits_{t,k} r(\hat{x}^{t_1}, \hat{x}^{k_1})\right)}{\max\limits_{t_1,k_1} r(\hat{x}^{t_1}, \hat{x}^{k_1}) - \min\limits_{t_1,k_1} r(\hat{x}^{t_1}, \hat{x}^{k_1})}. \tag{4.5}$$

   So, the matrix of correlation coefficients after the normalization can be treated as a matrix of a fuzzy tolerance relation. The D-PAFC-algorithm can then be applied directly to the matrix of normalized correlation coefficients;

3. The typical points of fuzzy clusters of the obtained principal allotment $R_P^{\alpha}(X)$ are selected as the most informative attributes.

Therefore, the idea of the described method is similar to the basic idea of the fuzzy attribute clustering method [20]. On the other hand, the step of the selection of the most informative attributes is simple in the proposed method. Moreover, the method can be considered as a version of the method of extremal grouping of attributes. An application of the method considered to solving the problem of the most informative attribute selection can be explained by a well-known example shown below.

4.2.2 *An Illustrative Example*

An application of the method considered to data visualization can be illustrated on Anderson's Iris data example. Let us consider the problem of the most informative attribute selection.

Anderson's Iris data forms the matrix of attributes $X_{4\times 150} = [\hat{x}_i^{t_1}]$, $i = 1,\ldots,150$, $t_1 = 1,\ldots,4$, where the sepal length is denoted by $\hat{x}^1$, the sepal width is denoted by $\hat{x}^2$, the petal length is denoted by $\hat{x}^3$ and the petal width is denoted by $\hat{x}^4$. The matrix of correlation coefficients $r_{4\times 4} = [r(\hat{x}^{t_1}, \hat{x}^{k_1})]$, $t_1, k_1 = 1,\ldots,4$, can be constructed using the formula (4.4) and is presented in Table 4.2.

Table 4.2 Matrix of the correlation coefficients for Anderson's Iris data

$r(\hat{x}^{t_1}, \hat{x}^{k_1})$	$\hat{x}^1$	$\hat{x}^2$	$\hat{x}^3$	$\hat{x}^4$
$\hat{x}^1$	1.0000	-0.1176	0.8718	0.8179
$\hat{x}^2$	-0.1176	1.0000	-0.4284	-0.3661
$\hat{x}^3$	0.8718	-0.4284	1.0000	0.9629
$\hat{x}^4$	0.8179	-0.3661	0.9629	1.0000

After application of the formula (4.5) to the matrix of correlation coefficients, the matrix of the fuzzy tolerance relation is constructed and is presented in Table 4.3.

Table 4.3 Matrix of the normalized correlation coefficients for Anderson's Iris data

$\tilde{r}(\hat{x}^{t_1}, \hat{x}^{k_1})$	$\hat{x}^1$	$\hat{x}^2$	$\hat{x}^3$	$\hat{x}^4$
$\hat{x}^1$	1.0000	0.2176	0.9102	0.8725
$\hat{x}^2$	0.2176	1.0000	0.0000	0.0436
$\hat{x}^3$	0.9102	0.0000	1.0000	0.9740
$\hat{x}^4$	0.8725	0.0436	0.9740	1.0000

The D-PAFC-algorithm is applied directly to the matrix of the normalized correlation coefficients. The principal allotment $R_P^{0.9102}(X)$ among two fuzzy clusters is obtained. The results obtained are presented in Table 4.4.

Table 4.4 Results obtained by the D-PAFC-algorithm: the attribute assignment

Class	Attributes			
	$\hat{x}^1$	$\hat{x}^2$	$\hat{x}^3$	$\hat{x}^4$
1	0.0000	1.0000	0.0000	0.0000
2	0.9102	0.0000	1.0000	0.9740

The second attribute $\hat{x}^2$ is the typical point of the first fuzzy cluster and the third attribute $\hat{x}^3$ is the typical point of the second fuzzy cluster. So, the attributes $\hat{x}^2$ and $\hat{x}^3$ can be selected as the most informative indexes.

Notice that the result is similar to the result obtained by using the conventional principal component analysis [102]. An interpretation of the obtained principal components can be made on the basis of factor loading. The factor loading is defined as a correlation coefficient between the ν-th principal component z_ν and the t_1-th attribute $\hat{x}^{t_1}$, $t_1 = 1, \ldots, m_1$, as follows [102]:

$$f(z_\nu, \hat{x}^{t_1}) = \frac{\operatorname{cov}\{z_\nu, \hat{x}^{t_1}\}}{\sqrt{V\{z_\nu\} V\{\hat{x}^{t_1}\}}}, \tag{4.6}$$

where $V\{z_\nu\}$ is the variance of z_ν, $V\{\hat{x}^{t_1}\}$ is the variance of $\hat{x}^{t_1}$, and $\operatorname{cov}\{z_\nu, \hat{x}^{t_1}\}$ is the covariance between z_ν and $\hat{x}^{t_1}$. In Table 4.5, each value shows the value of the factor loadings (4.6) which can show the relationship between each principal component and each attribute.

From the results obtained we can see how each component is explained by the attributes. This is related to the interpretation of each component. In Table 4.5, the first principal component is mainly explained by the attributes: sepal length, petal length, and petal width. Moreover, we can see a high correlation between the second principal component and the attribute of sepal width.

Table 4.5 Matrix of the normalized correlation coefficients for Anderson's Iris data

Attribute	Principal components			
	z_1	z_2	z_3	z_4
$\hat{x}^1$	0.89	0.36	0.28	0.04
$\hat{x}^2$	-0.46	0.88	-0.09	-0.02
$\hat{x}^3$	0.99	0.02	-0.05	-0.12
$\hat{x}^4$	0.96	0.06	-0.24	0.08

The two-dimensional projection of Anderson's Iris data can be constructed from the results obtained by the application of the proposed method. The projection is presented in Figure 4.2.

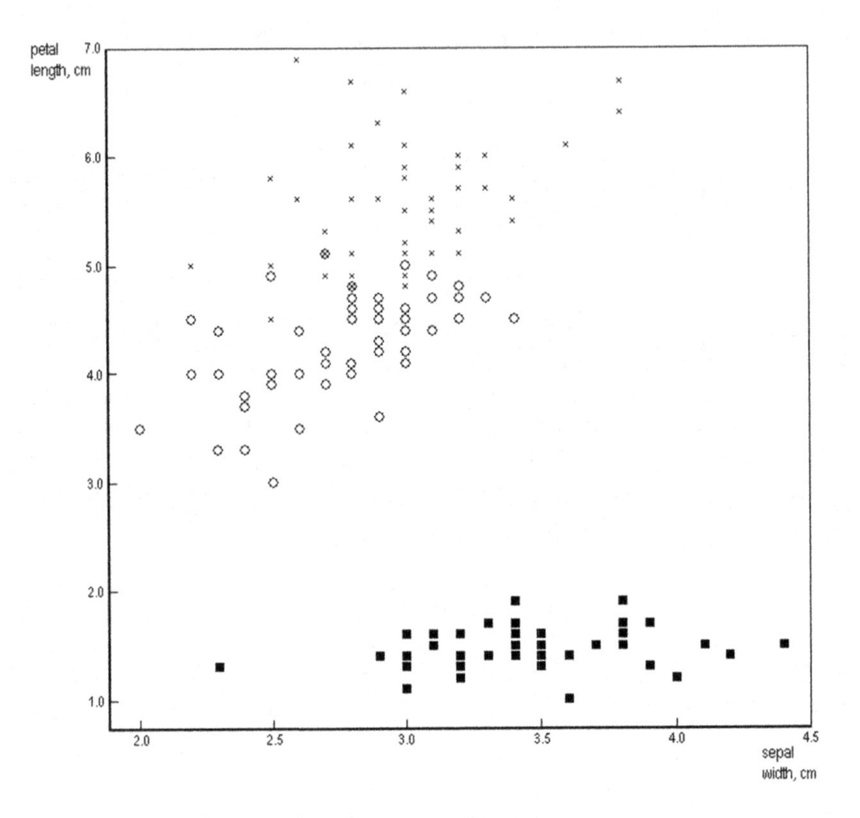

Fig. 4.2 Two-dimensional projection of Anderson's Iris data

Two well-separated classes are visualized. The first class corresponds to Iris Setosa. The second class corresponds to Iris Versicolor and Iris Virginica. The objects known to be Iris Setosa are represented by ■, while those known to be Iris Versicolor are represented by ○, and those known to be Iris Virginica are represented by ×. Obviously, this approach to data visualization can be very useful in exploratory data analysis.

By comparison of the results shown in Tables 4.4 and 4.5, we can see the following. In particular, values of the membership function of the first fuzzy cluster of the principal allotment $R_P^{0.9102}(X)$ can be interpreted as the normalized values of the factor loadings $f(z_2, \hat{x}^{t_1})$, $t_1 = 1,\ldots,4$, cf. Table 4.5 and the values of the membership function of the second fuzzy cluster of the principal allotment $R_P^{0.9102}(X)$ can be considered as the normalized values of the factor loadings $f(z_1, \hat{x}^{t_1})$, $t_1 = 1,\ldots,4$, cf. Table 4.5.

So, the result obtained by the application of the method presented to Anderson's Iris data shows that the proposed method is a simple, effective and efficient tool for the selection of the most relevant attributes, and that the method can be applied for data visualization. However, the proposed method is seemingly more appropriate a tool in comparison with the FACA-algorithm for solving the problem of attribute selection because the typical points of fuzzy clusters can be interpreted as the most relevant attributes and the typical points can be selected immediately. Moreover, the problem of cluster validity can directly be solved because the principal allotment among the fuzzy clusters is the classification result.

4.3 Clustering the Asymmetric Data

This section presents a technique of processing of the asymmetric data which is based on a heuristic algorithm of possibilistic clustering. This technique captures the transition of the asymmetric relational data into a fuzzy tolerance relation. The first subsection of the section provides a brief description of Sato's asymmetric fuzzy clustering model and a technique of the asymmetric data processing by relational heuristic algorithms of possibilistic clustering. In the second subsection an illustrative example is given.

4.3.1 A Technique for the Asymmetric Data Processing

The condition of symmetry of the initial data must be met in the relational approach to fuzzy clustering. However, proximity data can be asymmetric in some cases. For example, the communication in human relationships, information flow and the degree of confusion based on perception for a discrimination problem are some illustrative examples of asymmetric data. So, asymmetric data clustering methods play an important role in various real applications. That is why clustering techniques based on the asymmetric proximity have generated a great interest among a number of researchers. In particular, some issues of clustering of the objects for which asymmetric distances or proximities hold are studied by Owsiński [84].

On the other hand, an asymmetric clustering model considered by Sato-Ilic and Jain [102] is as follows:

$$\rho_{ij}^{A} = \sum_{l=1}^{c}\sum_{m=1}^{c} w_{lm} u_{li} u_{mj} + \varepsilon_{ij} , \tag{4.7}$$

where $\rho_{ij}^A = \rho^A(x_i, x_j)$, $0 \le \rho_{ij}^A \le 1$ is the observed asymmetric proximity degree between the objects x_i and x_j. The grade ρ_{ij}^A does not equal ρ_{ji}^A when $x_i \ne x_j$ and u_{li} is a value which represents the grade of membership of an object x_i to a fuzzy cluster A^l, $l \in \{1,\ldots,c\}$, under the condition (1.71). In the model (4.7), the weight w_{lm} is considered as a value which represents the asymmetric proximity between a pair of clusters A^l and A^m. So, the assumption that the asymmetry of proximity between the objects is caused by the asymmetry of proximity between the clusters is met in (4.7).

However, the clustering technique based on the model (4.7) is complex from the conceptual and mathematical points of view. On the other hand, if the observed data is obtained as an $n \times n$ asymmetric proximity matrix then this matrix can be interpreted as a preference relation on X. The matrix of a fuzzy preference relation can be obtained after the normalization to the matrix of initial data and a fuzzy tolerance relation can be obtained from a fuzzy preference relation. A technique serving those purposes was proposed in [145] can be described as follows:

1. The matrix of initial asymmetric data is normalized using the formula (2.20) and the matrix of normalized asymmetric dissimilarities or similarities $\rho_{n\times n}^A = [\rho_{ij}^A]$, $i, j = 1,\ldots,n$, is obtained;
2. The following condition is checked:

   if the matrix of the normalized asymmetric data $\rho_{n\times n}^A = [\rho_{ij}^A]$ satisfies the condition of antireflexivity (1.38),

   then the matrix of a weak fuzzy preference relation $R = [\mu_R(x_i, x_j)]$ is obtained by the application of the complementation operation (1.26) to the matrix $\rho_{n\times n}^A = [\rho_{ij}^A]$ and we go to step 3

   else the matrix $\rho_{n\times n}^A = [\rho_{ij}^A]$ of the weak fuzzy preference relation $R = [\mu_R(x_i, x_j)]$ is calculated and go to step 3;
3. The matrix of a fuzzy quasi-equivalence relation $Q = [\mu_Q(x_i, x_j)]$, $i, j = 1,\ldots,n$, can be obtained from the matrix of a weak fuzzy preference relation $R = [\mu_R(x_i, x_j)]$, $i, j = 1,\ldots,n$, according to the formula (1.51);
4. Some relational heuristic algorithm of possibilistic clustering is applied directly to the obtained matrix of a fuzzy quasi-equivalence relation Q.

The matrix of a fuzzy likeness relation L, which corresponds to the weak fuzzy preference relation R, is then be calculated in the Step 3 according to the formula (1.50) and the matrix $L=[\mu_L(x_i,x_j)]$, $i,j=1,\ldots,n$, is processed in Step 4. We will show below an example of the application of the proposed technique to the asymmetric data.

4.3.2 An Illustrative Example

Sato-Ilic and Jain's asymmetric data have originally appeared in [102]. This data shows human relations between 16 children. They show the degree of disliking between the children. So, the set of object will be denoted as $X=\{x_1,\ldots,x_{16}\}$ and the number of clusters c is determined as five. The original relation A is presented in Table 4.6.

Table 4.6 Dissimilarity matrix for the children

A	x_1	x_2	x_3	x_4	x_5	x_6	x_7	x_8	x_9	x_{10}	x_{11}	x_{12}	x_{13}	x_{14}	x_{15}	x_{16}
x_1	0	2	3	3	1	1	2	1	3	6	2	3	6	4	6	4
x_2	6	0	1	1	6	6	6	6	1	6	6	2	6	2	6	2
x_3	6	1	0	2	6	6	6	6	1	6	6	1	6	2	6	2
x_4	6	1	2	0	6	6	6	6	1	6	6	2	6	1	6	2
x_5	1	3	3	4	0	1	1	2	4	6	3	3	6	4	6	4
x_6	1	3	2	4	1	0	2	2	3	6	3	2	6	3	6	3
x_7	1	3	3	4	1	1	0	2	4	6	3	3	6	4	6	4
x_8	6	1	2	2	6	6	6	0	2	6	1	3	6	3	6	3
x_9	6	6	6	6	6	6	6	6	0	6	6	1	6	1	6	1
x_{10}	6	2	3	3	6	6	6	2	3	0	1	4	1	4	1	4
x_{11}	6	1	2	2	6	6	6	1	2	6	0	3	6	3	6	3
x_{12}	6	6	6	6	6	6	6	6	1	6	6	0	6	1	6	1
x_{13}	6	2	3	3	6	6	6	2	3	1	1	4	0	4	1	4
x_{14}	6	6	6	6	6	6	6	6	1	6	6	1	6	0	6	1
x_{15}	6	2	3	3	6	6	6	2	3	1	1	4	1	4	0	4
x_{16}	6	6	6	6	6	6	6	6	1	6	6	1	6	1	6	0

Let us consider the classification result using the model (4.7) which was presented by Sato-Ilic and Jain [102]. The fuzzy c-partition is the result obtained by using the model (4.7) to the asymmetric data. The membership functions u_{li} of five classes A^l, $l \in \{1,\ldots,5\}$, of the fuzzy c-partition are presented in Figure 4.3 and values which equal zero are not shown.

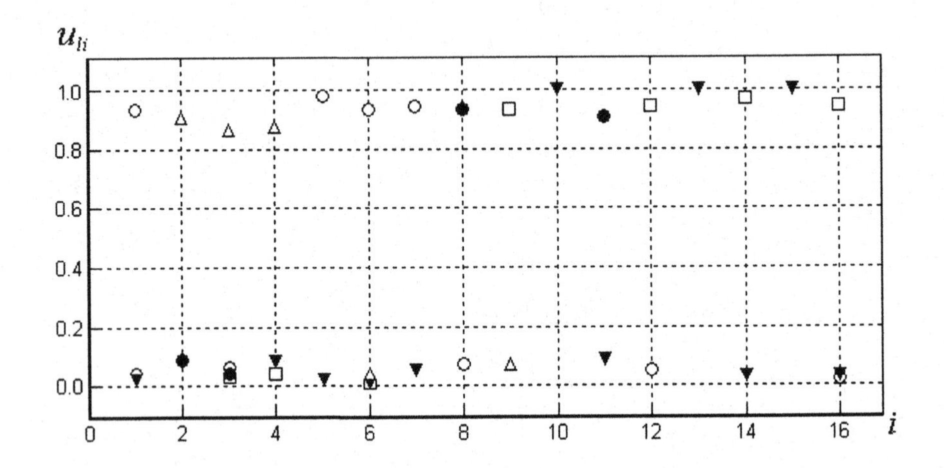

Fig. 4.3 Membership functions of five fuzzy clusters obtained from Sato-Ilic and Jain's classification method

The membership values of the first class are represented by ○, the membership values of the second class are represented by Δ, the membership values of the third class are represented by •, the membership values of the fourth class are represented by ▼, and the membership values of the fifth class are represented by □. Children who are in the same cluster are good friends with each other. Clusters A^1 and A^2 consist of boys only, and clusters A^3, A^4, A^5 include girls.

For comparison, the D-PAFC-algorithm was applied to the transformed data. Let us consider results of the experiments. A quasi-equivalence relation Q was obtained using the formula (1.51) in the first experiment. The corresponding matrix $Q = [\mu_Q(x_i, x_j)]$, $i, j = 1,\ldots,16$, was processed by the D-PAFC-algorithm and the principal allotment $R_P^{0.6666}(X)$ among five fully separate fuzzy clusters was obtained. The allotment, which corresponds to the result, was obtained for the tolerance threshold $\alpha = 0.6666$. The membership functions μ_{li} of five classes of the obtained allotment are presented in Figure 4.4.

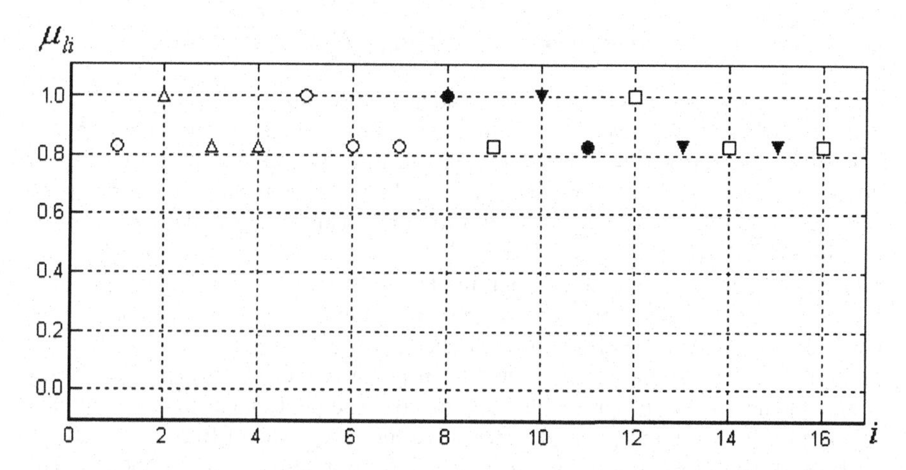

Fig. 4.4 Membership functions of five fuzzy clusters obtained from the D-PAFC-algorithm

The fifth object is the typical point of the first fuzzy cluster, the second object is the typical point of the fuzzy cluster which corresponds to the second class, the eighth object is the typical point of the fuzzy cluster which corresponds to the third class, the tenth object is the typical point of the fourth fuzzy cluster, and the twentieth object is the typical point of the fuzzy cluster which corresponds to the fifth class.

The fuzzy likeness relation L on X was constructed from the weak fuzzy preference relation R using the formula (1.50) in the second experiment and the corresponding matrix $L=[\mu_L(x_i,x_j)]$, $i,j=1,\ldots,16$, was processed by the D-PAFC-algorithm. The results obtained in the second experiment are equal to the results obtained in the first experiment for the quasi-equivalence relation Q. However, the resulting principal allotment $R_P^{0.8333}(X)$ was obtained for the value of tolerance threshold $\alpha=0.8333$.

So, the results which are obtained from the D-PAFC-algorithm using the proposed method of the asymmetric data preprocessing are similar to the results obtained by Sato-Ilic and Jain [102] using the model (4.7). Moreover, the membership function yielded by the proposed method is sharper than the membership function yielded by Sato-Ilic and Jain's method.

4.4 Discrimination of the Weak Fuzzy Preference Relations

This section describes a technique for constructing a subset of the most appropriate alternatives for the set of weak fuzzy preference relations which are defined on the universe of alternatives. In the first subsection methods of decision-making based on a fuzzy preference relation are described. In the second subsection the technique proposed is described in detail. The third subsection includes an example of application of the technique proposed to a data set.

4.4.1 Decision Making Based on Fuzzy Preference Relations

Fuzzy orderings were introduced by Zadeh [170]. Fuzzy orderings have been widely applied with success in information processing, decision-making and control, to just name a few areas. In general, a fuzzy ordering is a transitive or negatively transitive fuzzy binary relation. Fuzzy preference relations can be considered as a specific case of fuzzy orderings. This fact was demonstrated by Ovchinnikov [83].

A fuzzy subset of the most appropriate alternatives corresponds to a fuzzy preference relation and a problem of choosing a unique alternative of such a kind has been investigated by different researchers. However, a set of fuzzy preference relations can usually only be obtained from just a few decision makers. A fuzzy subset of the most appropriate alternatives corresponds to each fuzzy preference relation in this case and a unique alternative must be selected from the family of such fuzzy subsets. So, a unique fuzzy preference relation must be obtained and the construction of an appropriate decision is a fundamental problem in group decision theory. Many approaches can be used for this purpose. On the one hand, the construction of an appropriate consensus measure is very useful an approach which was considered by Kuzmin [70]. On the other hand, a unique fuzzy preference relation can be selected from the set of fuzzy preference relations proposed by members of a team. The problem was outlined by Nojiri [81].

An effective and efficient method of decision-making based on a fuzzy preference relation was proposed by Orlovsky [82]. A unique alternative can be selected from the set of alternatives using this method. We will outline it below.

Let $X = \{x_1, \ldots, x_n\}$ be a finite set of alternatives. Let R be some weak fuzzy preference relation on X in the sense of Definition 1.28 and $\mu_R(x_i, x_j)$, $\forall x_i, x_j \in X$, be its membership function. Let us consider the concept of strong fuzzy preference relation. The strong fuzzy preference relation is used for the construction of a fuzzy set of non-dominated alternatives. The strong fuzzy preference relation P which corresponds to the weak fuzzy preference relation R is a binary fuzzy relation on X with the membership function $\mu_P(x_i, x_j)$, $\forall x_i, x_j \in X$, defined as

$$\mu_P(x_i, x_j) = \begin{cases} \mu_R(x_i, x_j) - \mu_R(x_j, x_i), & if \quad \mu_R(x_i, x_j) > \mu_R(x_j, x_i) \\ 0, & if \quad \mu_R(x_i, x_j) \leq \mu_R(x_j, x_i) \end{cases}. \tag{4.8}$$

The fuzzy set $\tilde{R}$ of non-dominated alternatives is defined as follows:

$$\mu_{\tilde{R}}(x_i) = 1 - \sup_{x_j \in X} \mu_P(x_j, x_i), \forall x_i \in X. \tag{4.9}$$

So, the alternative $x_i \in X$ with the maximum values of the membership function $\max\limits_{x_i \in X} \mu_{\tilde{R}}(x_i)$ are the most appropriate alternatives.

Let us consider a method of constructing the fuzzy set of non-dominated alternatives for a set of weak fuzzy preference relations which are defined on a universe of alternatives. The method was described by Orlovsky [82].

Let $X = \{x_1, \ldots, x_n\}$ be a finite set of alternatives and $\{R^1, \ldots, R^g\}$ be a finite set of weak fuzzy preference relations on X. The fuzzy set of non-dominated alternatives can be constructed as $\tilde{R} = \tilde{R}_1 \cap \tilde{R}_2$, where $\tilde{R}_1$ is the fuzzy set of non-dominated alternatives which is determined in a set $(X, \mu_{R_1}(x_i, x_j))$, $\mu_{R_1}(x_i, x_j) = \min\{\mu_{R^1}(x_i, x_j), \ldots, \mu_{R^g}(x_i, x_j)\}$, $i, j = 1, \ldots, n$. On the other hand, $\tilde{R}_2$ is the fuzzy set of non-dominated alternatives which is determined in a set $(X, \mu_{R_2}(x_i, x_j))$, where R_2 is weighted aggregation of the weak fuzzy preference relations $R^1, \ldots, R^g$ which is described by

$$\mu_{R_2}(x_i, x_j) = \sum_{f=1}^{g} \lambda^f \mu_{R^f}(x_i, x_j), \tag{4.10}$$

and the condition

$$\sum_{f=1}^{g} \lambda^f = 1, \ \lambda^f \geq 0, \ f = 1, \ldots, g, \tag{4.11}$$

is met for all weights λ^f. If the values λ^f, are equal for all $f = 1, \ldots, g$, then $\lambda^f = 1/g$, $\forall f \in \{1, \ldots, g\}$.

A clustering technique for choosing a subset of the most appropriate weak fuzzy preference relations from the set of all weak fuzzy preference relations was proposed in [131] and the D-AFC(c)-algorithm of possibilistic clustering is the basis of this method. Let us consider its modification proposed in [131].

4.4.2 *A Clustering Approach to the Selection of a Unique Fuzzy Preference Relation*

Let us consider basic concepts of the technique in question as proposed in [131]. Let $X = \{x_1, \ldots, x_n\}$ be a finite set of alternatives and $\{R^1, \ldots, R^g\}$ be a finite set of weak fuzzy preference relations on X. By applying of the distance

$$d(R^k, R^f) = \sum_{(x_i, x_j)} \left| \mu_{R^k}(x_i, x_j) - \mu_{R^f}(x_i, x_j) \right|, k, f = 1, \ldots, g, \tag{4.12}$$

to the set $\{R^1, \ldots, R^g\}$ of weak fuzzy preference relations, a matrix of pair-wise dissimilarity coefficients $\rho_{g \times g} = [d(R^k, R^f)]$, $k, f = 1, \ldots, g$, is obtained. The distance (4.12) was proposed by Kuzmin [69], [70]. The matrix $\rho_{g \times g} = [d_{kf}]$ can be normalized using the formula (2.20). In fact, the matrix of normalized pair-wise dissimilarity coefficients $I = [\mu_I(R^k, R^f)]$, $k, f = 1,\ldots,g$, is the matrix of a fuzzy intolerance relation on the set of weak fuzzy preference relations $\{R^1,\ldots,R^g\}$. The matrix of fuzzy tolerance $T = [\mu_T(R^k, R^f)]$, $k, f = 1,\ldots,g$, is obtained by applying the complement operation

$$\mu_T(R^k, R^f) = 1 - \mu_I(R^k, R^f), \; k, f = 1,\ldots,g, \tag{4.13}$$

to the matrix of fuzzy intolerance $I = [\mu_I(R^k, R^f)]$. So, relational clustering algorithms can be applied directly to the matrix of fuzzy tolerance $T = [\mu_T(R^k, R^f)]$.

The basic idea of this approach is that the weak fuzzy preference relations can be classified and the typical points of each fuzzy cluster can be considered as elements of a subset of appropriate weak fuzzy preference relations. Elements of the subset of appropriate weak fuzzy preference relations must be ordered on the basis of their evaluations and an appropriate weak fuzzy preference relation must be selected.

A modified linear index of fuzziness can be used for the evaluation of the results of classification. A linear index of fuzziness was modified in [131] for the binary fuzzy relations as follows

$$\ell(R) = \frac{2}{n^2} \sum_{(x_i, x_j)} |\mu_R(x_i, x_j) - \mu_{\underline{R}}(x_i, x_j)|, \forall (x_i, x_j) \in X \times X, \tag{4.14}$$

where $n = card(X)$ is the number of alternatives and the crisp binary relation $\underline{R}$ is the nearest one to the fuzzy preference relation R.

The membership function $\mu_{\underline{R}}(x_i, x_j)$ of the crisp relation $\underline{R}$ can be defined as

$$\mu_{\underline{R}}(x_i, x_j) = \begin{cases} 0, & \mu_R(x_i, x_j) \le 0.5 \\ 1, & \mu_R(x_i, x_j) > 0.5 \end{cases}, \; \forall (x_i, x_j) \in R. \tag{4.15}$$

A unique fuzzy preference relation R can be selected if the condition $\min_f \ell(R^f)$, $f = 1,2,\ldots$, is met, where the fuzzy preference relations R^f are the typical points of fuzzy clusters obtained from some relational clustering algorithm.

Therefore, the method for solving the problem of selection of the most appropriate alternatives can be summarized as follows:

1. The matrix of fuzzy tolerance relation $T = [\mu_T(R^k, R^f)]$ is constructed according to the formulae (4.12), (2.20) and (4.13) for all weak fuzzy preference relations $R^1,\ldots,R^g$;
2. The D-PAFC-algorithm is applied directly to the matrix of the fuzzy tolerance relation $T = [\mu_T(R^k, R^f)]$;
3. The typical points of the fuzzy clusters of the obtained principal allotment $R_P^{\alpha}(X) = \{A_{(\alpha)}^l \mid l = \overline{1,c}\}$ among the unknown least number of fully separate fuzzy clusters are selected as the elements of a subset of appropriate weak fuzzy preference relations;
4. The value of the modified linear index of fuzziness (4.14) is calculated for each element of the subset of appropriate weak fuzzy preference relations;
5. The following condition is checked:

   if for some unique fuzzy preference relation R the condition $\min_f \ell(R^f)$, $f = 1,2,\ldots$ is met,

   then the fuzzy preference relation R can be selected as a unique element of the subset of the most appropriate weak fuzzy preference relations

   else the condition $\min_f \ell(R^f)$, $f = 1,2,\ldots$, is met for several fuzzy preference relation R which are elements of that subset;
6. The fuzzy set of of non-dominated alternatives is determined for each fuzzy preference relation R which is an element of the subset of the most appropriate weak fuzzy preference relations and alternatives $x_i \in X$ with the maximum values of the membership function $\max_{x_i \in X} \mu_{\tilde{R}}(x_i)$ are elements of the subset of the most appropriate alternatives.

The subset of the most appropriate alternatives results therefore from the application of the described procedure to the set of weak fuzzy preference relations $\{R^1,\ldots,R^g\}$ defined on the set of alternatives $X = \{x_1,\ldots,x_n\}$.

4.4.3 An Illustrative Example

The application of the technique presented for solving the problem of selection of the most appropriate alternative can be explained by an illustrative example [131]. First, the description of the data set should be provided. Let $X = \{x_1, x_2, x_3, x_4\}$ be a finite set of alternatives. Let $\{R^1, R^2, R^3, R^4, R^5, R^6\}$ be a set of weak fuzzy preference relations on X. The set of fuzzy preference relations is portrayed below in Figure 4.5.

R^1	x_1	x_2	x_3	x_4
x_1	1.0	0.2	0.3	0.1
x_2	0.5	1.0	0.2	0.6
x_3	0.1	0.6	1.0	0.3
x_4	0.6	0.1	0.5	1.0

R^2	x_1	x_2	x_3	x_4
x_1	1.0	0.3	0.4	0.2
x_2	0.6	1.0	0.3	0.7
x_3	0.2	0.7	1.0	0.4
x_4	0.7	0.2	0.6	1.0

R^3	x_1	x_2	x_3	x_4
x_1	1.0	0.3	0.4	0.2
x_2	0.5	1.0	0.3	0.7
x_3	0.1	0.6	1.0	0.4
x_4	0.6	0.1	0.5	1.0

R^4	x_1	x_2	x_3	x_4
x_1	1.0	0.4	0.1	0.3
x_2	0.2	1.0	0.4	0.4
x_3	0.3	0.4	1.0	0.5
x_4	0.4	0.3	0.3	1.0

R^5	x_1	x_2	x_3	x_4
x_1	1.0	0.5	0.2	0.4
x_2	0.1	1.0	0.5	0.5
x_3	0.2	0.3	1.0	0.6
x_4	0.3	0.2	0.2	1.0

R^6	x_1	x_2	x_3	x_4
x_1	1.0	0.4	0.1	0.3
x_2	0.3	1.0	0.4	0.4
x_3	0.4	0.5	1.0	0.5
x_4	0.5	0.4	0.4	1.0

Fig. 4.5 Six weak fuzzy preference relations

The set of objects for classification constitutes therefore six weak fuzzy preference relations. The matrix of pair-wise dissimilarity coefficients $\rho_{g \times g} = [d_{kf}]$, $k, f = 1, \ldots, 6$, is obtained by the application of the formula (4.12) to these six weak fuzzy preference relations. The matrix of pair-wise dissimilarity coefficients is presented in Table 4.7.

Table 4.7 The matrix of pair-wise dissimilarity coefficients for the data set

d_{kf}	R^1	R^2	R^3	R^4	R^5	R^6
R^1	0.0					
R^2	1.2	0.0				
R^3	0.6	0.6	0.0			
R^4	2.5	2.5	2.3	0.0		
R^5	2.9	2.9	2.7	1.2	0.0	
R^6	2.3	2.3	2.1	0.6	1.8	0.0

The matrix of pair-wise dissimilarity coefficients is normalized using the formula (2.20). The matrix of fuzzy tolerance relation is constructed for these six weak fuzzy preference relations by applying the formula (4.13) to the matrix of normalized pair-wise dissimilarity coefficients $I = [\mu_I(R^k, R^f)]$, $k, f = 1,\ldots,6$. By executing the D-PAFC-algorithm, we obtain the following: the first class is formed by 3 elements, and the second class is composed of 3 elements too. The membership functions of two classes are presented in Figure 4.6.

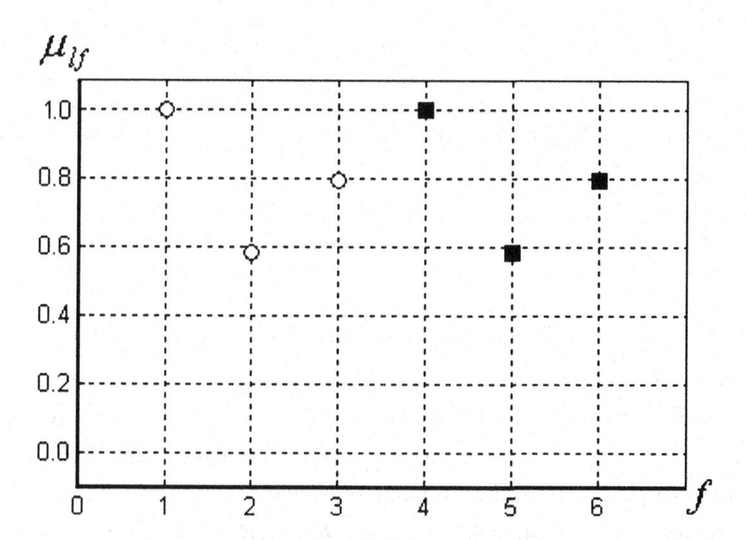

Fig. 4.6 Membership functions of two fuzzy clusters obtained from the D-PAFC-algorithm

The membership values of the first class are represented in Figure 4.6 by ○ and the membership values of the second class are represented by ■. The value of the membership function of the fuzzy cluster which corresponds to the first class is maximal for the first object and is equal one. So, the first object is the typical point of the fuzzy cluster which corresponds to the first class. The membership value of the fourth object is equal one for the fuzzy cluster which corresponds to the second class. Thus, the fourth object is the typical point of the fuzzy cluster which corresponds to the second class. That is why the set $\{R^1, R^4\}$ is the subset of the appropriate weak fuzzy preference relations. Obviously, the analysis of the set of weak fuzzy preference relations is a simple problem.

The values of the modified linear index of fuzziness (4.14) for both elements of the set $\{R^1, R^4\}$ are calculated. The values $\ell(R^1) = 0.425$ and $\ell(R^4) = 0.500$ are obtained. So, the fuzzy preference relation R^1 can be selected as the unique element of the subset of the most appropriate weak fuzzy preference relations from the set $\{R^1,\ldots,R^6\}$ of weak fuzzy preference relations on X.

A fuzzy set $\tilde{R}^1$ of non-dominated alternatives can be constructed as follows:

$$\tilde{R}^1 = ((x_1, 0.5), (x_2, 0.6), (x_3, 0.8), (x_4, 0.5)). \tag{4.14}$$

So, the alternative x_3 can be selected as the most appropriate alternative from the set $X = \{x_1, x_2, x_3, x_4\}$ and the membership degree $\mu_{\tilde{R}^1}(x_3) = 0.8$ is a grade of acceptance of the alternative x_3.

However, the method should be compared with other methods. Let us consider the results obtained from both Orlovsky's methods. On the one hand, the method which is based on the weighted aggregation of the weak fuzzy preference relations (4.10) and (4.11) is applied directly to the set of weak fuzzy preference relations $\{R^1, \ldots, R^6\}$ with values $\lambda^f = 1/6$, for all $f = 1, \ldots, 6$. So, a fuzzy set $\tilde{R}$ of non-dominated alternatives is constructed and can be written as follows:

$$\tilde{R} = ((x_1, 0.73), (x_2, 0.83), (x_3, 0.97), (x_4, 0.67)). \tag{4.15}$$

That is why the alternative x_3 can be selected as the most appropriate alternative from the set $X = \{x_1, x_2, x_3, x_4\}$.

On the other hand, one should consider results of constructing the fuzzy set of non-dominated alternatives for each weak fuzzy preference relation $\{R^1, \ldots, R^6\}$. Thus, the method based on the formulae (4.8) and (4.9) is applied for each weak fuzzy preference relation from the set $\{R^1, \ldots, R^6\}$. So, six strong fuzzy preference relations P^f, $f = 1, \ldots, 6$, and six fuzzy sets of non-dominated alternatives $\tilde{R}^f$, $f = 1, \ldots, 6$, are constructed. The fuzzy sets of non-dominated alternatives $\tilde{R}^f$, $f = 1, \ldots, 6$, are presented in Table 4.8.

Table 4.8 Fuzzy sets of non-dominated alternatives

Alternatives	Fuzzy sets of non-dominated alternatives					
	$\tilde{R}^1$	$\tilde{R}^2$	$\tilde{R}^3$	$\tilde{R}^4$	$\tilde{R}^5$	$\tilde{R}^6$
x_1	0.5	0.5	0.6	0.8	1.0	0.7
x_2	0.6	0.6	0.7	0.8	0.6	0.9
x_3	0.8	0.8	0.7	1.0	0.8	1.0
x_4	0.5	0.5	0.4	0.8	0.6	0.9

The result obtained shows that the fuzzy set of non-dominated alternatives $\tilde{R}^6$ is the most appropriate fuzzy set for obvious reasons. Obviously, the alternative x_3 can be selected as the most appropriate alternative for the fuzzy sets of non-dominated alternatives $\tilde{R}^1$, $\tilde{R}^2$, $\tilde{R}^4$, and $\tilde{R}^6$. The alternative x_1 can be selected as the most appropriate alternative for the fuzzy set of non-dominated alternatives $\tilde{R}^5$. The alternatives x_2 and x_3 can be selected as two most appropriate alternatives for the fuzzy set of non-dominated alternatives $\tilde{R}^3$.

Let us consider the values of the modified linear index of fuzziness (4.14) for all weak fuzzy preference relations. These values are presented in Table 4.9.

Table 4.9 Values of the modified linear index of fuzziness for weak fuzzy preference relations

Fuzzy relations	R^1	R^2	R^3	R^4	R^5	R^6
Value of the index (4.14)	0.4250	0.4625	0.4875	0.5000	0.4750	0.5750

So, the condition $\min_f \ell(R^f)$, $f = 1,\ldots,6$, is met for the relation R^1 and this corresponds to the result of the evaluation of the weak fuzzy preference relations R^1 and R^4 which are the typical points of fuzzy clusters. Notably, the values of the modified linear index of fuzziness (4.14) do not corresponds to the results which are presented in Table 4.8. For example, the fuzzy set $\tilde{R}^1$ is equal to the fuzzy set $\tilde{R}^2$ and $\ell(R^1) < \ell(R^2)$. So, the same fuzzy sets of non-dominated alternatives can be obtained from different weak fuzzy preference relations. Moreover, the classification results do not depend on values of the modified linear index of fuzziness (4.14). For example, the fuzzy relation R^4 is the typical point of the second class and the membership function values are ordered as follows $\mu_{24} > \mu_{26} > \mu_{25}$ for the elements of the second fuzzy cluster. However, the values of the index (4.14) are ordered as follows $\ell(R^5) < \ell(R^4) < \ell(R^6)$ for fuzzy relations which are elements of the corresponding class.

The application of the D-PAFC-algorithm to the classification of weak fuzzy preference relations is compared with the relational ARCA-algorithm of fuzzy clustering [25]. For this purpose, the objective function of the model can be rewritten as follows:

$$Q_{ARCA}(P,\mathrm{T}) = \sum_{l=1}^{c}\sum_{k=1}^{g} u_{lk}^{\gamma}\left(\sqrt{\sum_{f=1}^{g}\left(\mu_I(R^k,R^f) - d(R^f,\tau^l)\right)^2}\right)^2 . \qquad (4.16)$$

where $\mu_I(R^k,R^f)$ is the fuzzy intolerance relation between the pair of weak fuzzy preference relations R^k and R^f, and $d(R^f,\tau^l)$ is the relation between the prototype τ^l, $l \in \{1,\ldots,c\}$, and the object R^f, $f \in \{1,\ldots,g\}$.

So, the ARCA-algorithm is applied directly to the matrix of the fuzzy intolerance relation $I = [\mu_I(R^k, R^f)]$, $k, f = 1,\ldots,6$, for $c = 2$ and $\gamma = 2$. The matrix $P_{2\times 6} = [u_{lf}]$, $l = 1,2$, $f = 1,\ldots,6$, of the fuzzy c-partition $P(X)$ is obtained and the membership functions of two classes are presented in Figure 4.7. The membership values of the first class are represented by ○ and the membership values of the second class are represented by ■.

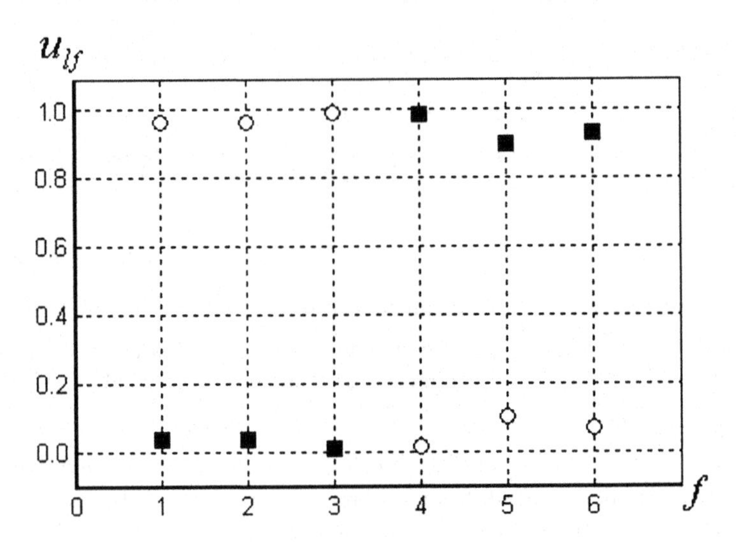

Fig. 4.7 Membership functions of two fuzzy clusters obtained from the ARCA-algorithm

On the other hand, the matrix of dissimilarities $K_{c\times n} = [d_{lf}]$ between the prototype τ^l, $l \in \{1,2\}$ and the fuzzy relation R^f, $f \in \{1,\ldots,6\}$, is also a result of classification. The values of dissimilarity coefficients $d(R^f, \tau^l)$, $f = 1,\ldots,6$, for both classes $A^l \in P(X)$, $l = 1,2$, are shown in Figure 4.8. The values of dissimilarity coefficients $d(R^f, \tau^l)$ for the first class objects are represented by ○ and the values of dissimilarities $d(R^f, \tau^l)$ for the second class objects are represented by ■.

The value of the membership function u_{lf}, $l = 1,2$, $f = 1,\ldots,6$, of the fuzzy cluster which corresponds to the first class is maximal for the third object, R^3. The membership value of the fourth object R^4 is maximal for the fuzzy cluster which corresponds to the second class. On the other hand, the value of the dissimilarity coefficient $d(R^f, \tau^l)$ is minimal for the third object R^3 and the first class prototype, τ^1. The value of the dissimilarity coefficient $d(R^f, \tau^l)$ is

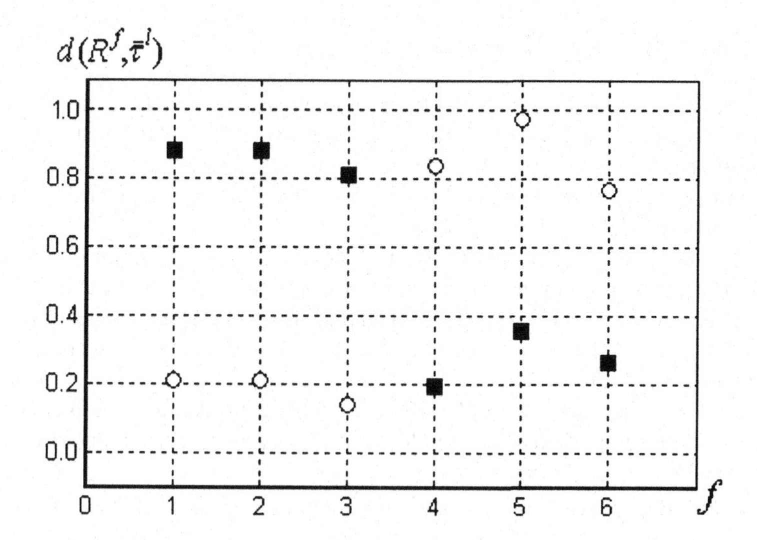

Fig. 4.8 Values of dissimilarities between prototypes of two fuzzy clusters and objects

minimal for the fourth object R^4 and the second class prototype, τ^2. So, the weak fuzzy preference relations R^3 and R^4 can be selected as elements of the subset of appropriate weak fuzzy preference relations. The condition $\ell(R^3) < \ell(R^4)$ is met and the fuzzy set of non-dominated alternatives $\tilde{R}^3$ is constructed for the fuzzy relation R^3. That is why the alternatives x_2 and x_3 are selected as two most appropriate alternatives. So, the application of the ARCA-algorithm for constructing the subset of appropriate weak fuzzy preference relations is not appropriate in the considered example.

In general, the considered technique for the classification of weak fuzzy preference relations seems to be satisfactory for the construction of the subset of the most appropriate weak fuzzy preference relations for a detailed analysis.

4.5 Automatic Generation of a Fuzzy Inference Systems from Data

The problem of extracting fuzzy rules from data is considered in the present section. The first subsection includes a discussion of methods for extracting fuzzy rules based on fuzzy clustering. The second subsection includes a technique of fuzzy inference system prototyping based on the clustering results obtained from the D-AFC(c)-algorithm. Illustrative examples are presented in the third subsection.

4.5.1 Preliminary Remarks

Fuzzy inference systems are one of the prominent and visible applications of fuzzy logic and fuzzy sets theory. They can be useful in classification, process simulation and diagnosis, online decision support, process control, etc. So, the problem of generation of fuzzy rules is one of more important problems in the development of fuzzy inference systems.

There are a number of approaches to learning fuzzy rules from data based on the techniques of evolutionary or neural computation, mostly aiming at optimizing parameters of fuzzy rules. Fuzzy clustering seems to be a very appealing method for learning fuzzy rules since there is a close and canonical connection between fuzzy clusters and fuzzy rules.

The idea of deriving fuzzy classification rules from data can be formulated as follows: the training data set is divided into homogeneous group and a fuzzy rule is associated to each group.

The training set contains n data pairs. Each pair is made of a m_1-dimensional input-vector and a c-dimensional output-vector. We assume that the number of rules in the fuzzy inference system rule base is c.

Mamdani's [75] rule l in the fuzzy inference system is written as follows:

$$If\ \hat{x}^1\ is\ B_l^1\ and \ldots and\ \hat{x}^{m_1}\ is\ B_l^{m_1}\ then\ y_1\ is\ C_1^l\ and \ldots and\ y_c\ is\ C_c^l, \quad (4.17)$$

where $B_l^{t_1}$, $t_1 \in \{1,\ldots,m_1\}$ and C_l^l, $l \in \{1,\ldots,c\}$, are fuzzy sets that define an input and output space partitioning.

A fuzzy inference system which is described by a set of fuzzy rules of the form (4.17) is a MIMO (multiple input, multiple output) system. Note that any fuzzy rule of the form (4.17) can be represented by c rules of the form of MISO (multiple input, single output) system:

$$\begin{array}{l} If\ \hat{x}^1\ is\ B_l^1\ and \ldots and\ \hat{x}^{m_1}\ is\ B_l^{m_1}\ then\ y_1\ is\ C_1^l \\ \ldots \\ If\ \hat{x}^1\ is\ B_l^1\ and \ldots and\ \hat{x}^{m_1}\ is\ B_l^{m_1}\ then\ y_c\ is\ C_c^l \end{array}. \quad (4.18)$$

Let $B_l^{t_1}$ be characterized by the membership function $\gamma_{B_l^{t_1}}(\hat{x}^{t_1})$. The membership function can be triangular, Gaussian, trapezoidal, or any other shape. Fuzzy classification rules can be obtained directly from fuzzy clustering results. The principal idea of extracting fuzzy classification rules based on fuzzy clustering is the following [49]. Each fuzzy cluster is assumed to be assigned to one class for classification and the membership grades of the data to the clusters determine the degree to which they can be classified as a member of the corresponding class. So, with a fuzzy cluster that is assigned to the some class we can associate a linguistic rule. The fuzzy cluster is projected into each single dimension leading to a fuzzy set on the real numbers. An approximation of the fuzzy set by projecting only the data set and computing the convex hull of this projected fuzzy set or approximating it by a trapezoidal or triangular membership function is used to obtain the rules [49].

The objective function-based fuzzy clustering algorithms are sensitive to the selection of the initial partition and the fuzzy rules depend on the selection of the fuzzy clustering method. In particular, the GG-algorithm and the GK-algorithm of fuzzy clustering are recommended in [49] for the generation of fuzzy rules.

4.5.2 Extracting Fuzzy Rules from Fuzzy Clusters

Let us consider a method of extracting fuzzy classification rules based on a heuristic method of possibilistic clustering which was outlined in [142]. In the following, we will assume that the Mamdani-type fuzzy inference system is a MIMO (multiple input, multiple output) system.

The antecedent of a fuzzy rule in a fuzzy inference system defines a decision region in the m_1-dimensional feature space. Let us consider a fuzzy rule (4.17) where $B_l^{t_1}$, $t_1 = 1,\ldots,m_1$, $l \in \{1,\ldots,c\}$ is a fuzzy set associated with the attribute variable $\hat{x}^{t_1}$. Let $B_l^{t_1}$ be characterized by the trapezoidal membership function $\gamma_{B_l^{t_1}}(\hat{x}^{t_1})$ which is shown in Figure 4.9.

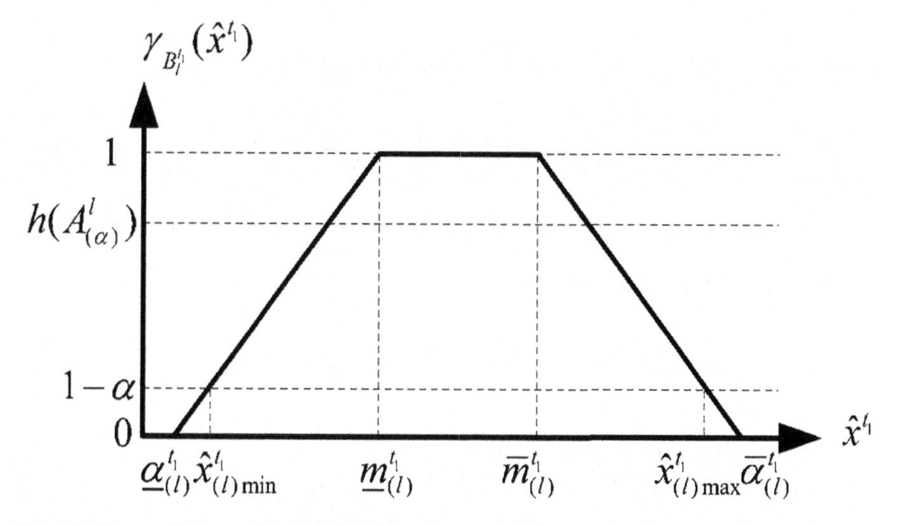

Fig. 4.9 Trapezoidal membership function of an antecedent fuzzy set

So, the fuzzy set $B_l^{t_1}$ can be defined by four parameters, $B_l^{t_1} = (\underline{a}^{t_1}_{(l)}, \underline{m}^{t_1}_{(l)}, \overline{m}^{t_1}_{(l)}, \overline{a}^{t_1}_{(l)})$. A triangular fuzzy set $B_l^{t_1} = (\underline{a}^{t_1}_{(l)}, m^{t_1}_{(l)}, \overline{a}^{t_1}_{(l)})$ can be considered as a particular case of the trapezoidal fuzzy set where $\underline{m}^{t_1}_{(l)} = \overline{m}^{t_1}_{(l)}$.

The idea of deriving fuzzy rules from fuzzy clusters is the following [142]. We apply the D-AFC(c)-algorithm to the given data and then obtain for each fuzzy cluster $A^l_{(\alpha)}$, $l \in \{1,\ldots,c\}$ a kernel $K(A^l_{(\alpha)})$ and a support A^l_α. The value of a tolerance threshold $\alpha \in (0,1]$, which corresponds to the allotment $R^*_c(X) = \{A^1_{(\alpha)},\ldots,A^c_{(\alpha)}\}$, is the additional result of classification. We calculate the interval $[\hat{x}^{t_1}_{(l)\min}, \hat{x}^{t_1}_{(l)\max}]$ of values of each attribute $\hat{x}^{t_1}$, $t_1 \in \{1,\ldots,m_1\}$, for the support A^l_α. The value $\hat{x}^{t_1}_{(l)\min}$ can be obtained as follows

$$\hat{x}^{t_1}_{(l)\min} = \min_{x_i \in A^l_\alpha} \hat{x}^{t_1}, \ \forall t_1 \in \{1,\ldots,m_1\}, \ \forall l \in \{1,\ldots,c\}, \tag{4.19}$$

and the value $\hat{x}^{t_1}_{(l)\max}$, $t_1 \in \{1,\ldots,m_1\}$ can be calculated using the formula

$$\hat{x}^{t_1}_{(l)\max} = \max_{x_i \in A^l_\alpha} \hat{x}^{t_1}, \ \forall t_1 \in \{1,\ldots,m\}, \ \forall l \in \{1,\ldots,c\}. \tag{4.20}$$

The parameter $\underline{a}^{t_1}_{(l)}$ can be obtained as follows:

$$\gamma_{B^{t_1}_l}(\hat{x}^{t_1}_{(l)\min}) = (1-\alpha), \ \gamma_{B^{t_1}_l}(\underline{a}^{t_1}_{(l)}) = 0, \tag{4.21}$$

and the parameter $\overline{a}^{t_1}_{(l)}$ can be obtained from the conditions:

$$\gamma_{B^{t_1}_l}(\hat{x}^{t_1}_{(l)\max}) = (1-\alpha), \ \gamma_{B^{t_1}_l}(\overline{a}^{t_1}_{(l)}) = 0. \tag{4.22}$$

We calculate the value $\underline{\hat{x}}^{t_1}_{(l)}$ for all typical points $\tau^l_e \in K(A^l_{(\alpha)})$ of the fuzzy cluster $A^l_{(\alpha)}$, $l \in \{1,\ldots,c\}$ as follows:

$$\underline{\hat{x}}^{t_1}_{(l)} = \min_{\tau^l_e \in K(A^l_{(\alpha)})} \hat{x}^{t_1}, \ \forall e \in \{1,\ldots,|l|\}, \tag{4.23}$$

and the value $\overline{\hat{x}}^{t_1}_{(l)}$ can be obtained from

$$\overline{\hat{x}}^{t_1}_{(l)} = \max_{\tau^l_e \in K(A^l_{(\alpha)})} \hat{x}^{t_1}, \ \forall e \in \{1,\ldots,|l|\}. \tag{4.24}$$

Thus, the parameter $\underline{m}^{t_1}_{(l)}$ can be calculated from the conditions:

$$\gamma_{B^{t_1}_l}(\underline{\hat{x}}^{t_1}_{(l)}) = \gamma_{B^{t_1}_l}(\underline{m}^{t_1}_{(l)}) = 1, \tag{4.25}$$

and the parameter $\overline{m}_{(l)}^{t_1}$ can be obtained as follows:

$$\gamma_{B_l^{t_1}}(\hat{x}_{(l)}^{t_1}) = \gamma_{B_l^{t_1}}(\overline{m}_{(l)}^{t_1}) = 1. \quad (4.26)$$

The height $h(A_{(\alpha)}^l) = \sup_{x_i \in A_\alpha^l} \mu_{A_{(\alpha)}^l}(x_i)$ of the fuzzy cluster $A_{(\alpha)}^l$, $l \in \{1,\ldots,c\}$, must be taken into account because the fuzzy cluster $A_{(\alpha)}^l \in R_c^*(X)$ can be a subnormal fuzzy set [133], [134].

So, the condition $\check{x}_{(l)}^{t_1} = \underline{m}_{(l)}^{t_1}$ and the condition $\hat{x}_{(l)}^{t_1} = \overline{m}_{(l)}^{t_1}$ are met for all input variables $\hat{x}^{t_1}$, $t_1 = 1,\ldots,m_1$. Obviously, if the condition $h(A_{(\alpha)}^l) = 1$ is met for the fuzzy cluster $A_{(\alpha)}^l$ and only one typical point is present in the fuzzy cluster, then the condition $\check{x}_{(l)}^{t_1} = \hat{x}_{(l)}^{t_1} = \underline{m}_{(l)}^{t_1} = \overline{m}_{(l)}^{t_1}$ is met.

Let us consider a technique of learning the consequents of the rules. The variables y_l, $l = 1,\ldots,c$, are the consequents of the fuzzy rules (4.17), represented by the fuzzy sets C_l^l, $l = 1,\ldots,c$, with the membership functions $\gamma_{C_l^l}(y_l)$. The fuzzy sets C_l^l, $l = 1,\ldots,c$, can be defined on the interval of memberships [0,1] and these fuzzy sets can be presented as follows: $C_l^l = (\alpha, \underline{\mu}_l, \overline{\mu}_l, 1)$, where α is a tolerance threshold, $\underline{\mu}_l = \min_{x_i \in A_\alpha^l} \mu_{li}$ and $\overline{\mu}_l = \max_{x_i \in A_\alpha^l} \mu_{li}$. The membership function can be interpreted as a high membership. On the other hand, if $A_{(\alpha)}^l$ and $A_{(\alpha)}^m$, $l \neq m$, are two particular separated fuzzy clusters, then the condition $w \neq 0$ is met in the equation (2.7). So, the fuzzy set $C_m^l = (0, 1-\overline{\mu}_m, 1-\underline{\mu}_m, 1-\alpha)$ is the consequent for the variable y_m of the l-th fuzzy rule for the case of a low membership.

So, the membership functions $\gamma_{C_l^l}(y_l)$ of the fuzzy sets C_l^l, $l = 1,\ldots,c$, will be the trapezoidal membership functions. The fuzzy clusters can be subnormal fuzzy sets [139]. The cases which are presented in Figures 4.10 – 4.11 are therefore general cases.

If the allotment $R_c^*(X)$ among fully separate fuzzy clusters is obtained and all fuzzy clusters $A_{(\alpha)}^l \in R_c^*(X)$, $l \in \{1,\ldots,c\}$, are normal fuzzy sets, then $h(A_{(\alpha)}^l) = 1$, $\forall A_{(\alpha)}^l \in R_c^*(X)$, and $\overline{\mu}_l = 1$ for each fuzzy cluster $A_{(\alpha)}^l$, $l \in \{1,\ldots,c\}$.

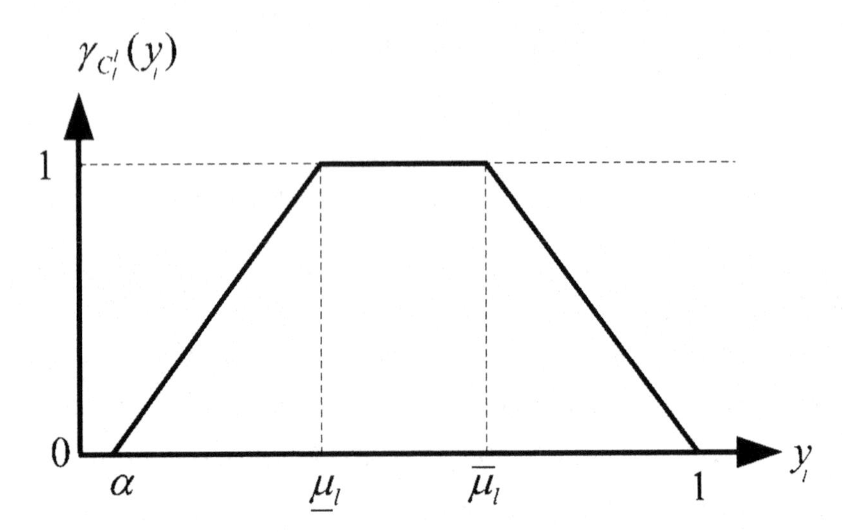

Fig. 4.10 Membership function for a consequent fuzzy set in the case of a high degree of belongingness

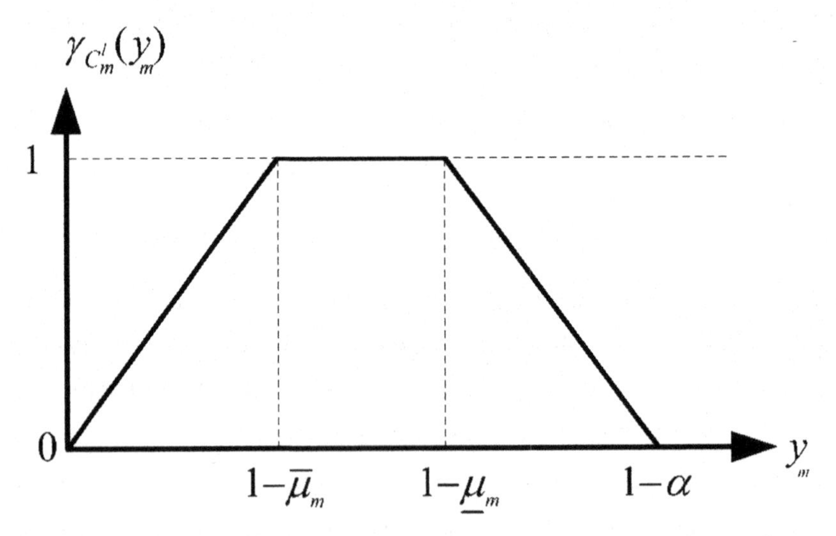

Fig. 4.11 Membership function for a consequent fuzzy set in the case of a low degree of belongingess

Thus, the trapezoidal membership functions $\gamma_{C_l^l}(y_l)$ for the fuzzy sets C_l^l, $l = 1,\ldots,c$, can be constructed on the basis of the clustering results. The empty set $A_\alpha^l = \varnothing$, $l \in \{1,\ldots,c\}$, can correspond to some output variable y_l, $l \in \{1,\ldots,c\}$. So, the empty fuzzy set C_l^l will correspond to the output variable

y_l, $l \in \{1,\ldots,c\}$ and $\gamma_{C_l^l}(y_l)=0$ is the membership function of the corresponding fuzzy set C_l^l.

The accuracy threshold ε can be useful for tuning the rules. In particular, for $\varepsilon \to 0$ we have $(1-\alpha) \to 0$ and the crisp interval $[\underline{a}_{(l)}^{t_1}, \overline{a}_{(l)}^{t_1}]$ increases. Otherwise, if we decrease the accuracy threshold, $\varepsilon \to 1$, then the number of typical points $|l|$ of the fuzzy cluster $A_{(\alpha)}^l \in R_c^*(X)$ increases, and the crisp interval $[\underline{m}_{(l)}^{t_1}, \overline{m}_{(l)}^{t_1}]$ increases.

The membership functions $\gamma_{C_l^l}(y_l)$ for the consequent fuzzy sets C_l^l, $l=1,\ldots,c$, depend on the value of the accuracy threshold ε. For example, if we increase the value of the accuracy threshold ε, the crisp interval $[\alpha,1]$ decreases. Moreover, for $\varepsilon \to 1$ we have $\mu_{li} \to 1$, $l=1,\ldots,c$, $i=1,\ldots,n$. That is why the parameters $\underline{\mu}_l$ and $\overline{\mu}_l$ increase for all fuzzy sets C_l^l, i.e., for $\varepsilon \to 1$, we have $\underline{\mu}_l \to 1$ and $\overline{\mu}_l \to 1$, $\forall l=1,\ldots,c$.

The described method of constructing fuzzy inference systems can be used for extracting fuzzy classification rules from the interval-valued training data set. This fact was shown in [143]. The case of the interval-valued data can be described by the expression $\hat{x}_i^{t_1} = (\hat{x}_i^{t_1(\min)}, \hat{x}_i^{t_1(\max)})$, $t_1=1,\ldots,m_1$, $i=1,\ldots,n$. In particular, the interval $[\hat{x}_{(l)\min}^{t_1(\min)}, \hat{x}_{(l)\max}^{t_1(\min)}]$ of values of each attribute $\hat{x}^{t_1} = (\hat{x}^{t_1(\min)}, \hat{x}^{t_1(\max)})$, $t_1 \in \{1,\ldots,m_1\}$, for the support A_α^l should be calculated. The value $\hat{x}_{(l)\min}^{t_1(\min)}$ can be obtained as follows:

$$\hat{x}_{(l)\min}^{t_1(\min)} = \min_{x_i \in A_\alpha^l} \hat{x}^{t_1(\min)}, \ \forall t_1 \in \{1,\ldots,m_1\}, \ \forall l \in \{1,\ldots,c\}, \tag{4.27}$$

and the value $\hat{x}_{(l)\max}^{t_1(\max)}$, $t_1 \in \{1,\ldots,m_1\}$ can be calculated using the rewritten formula (4.20)

$$\hat{x}_{(l)\max}^{t_1(\max)} = \max_{x_i \in A_\alpha^l} \hat{x}^{t_1(\max)}, \ \forall t \in \{1,\ldots,m\}, \ \forall l \in \{1,\ldots,c\}. \tag{4.28}$$

Thus, the parameters $\underline{a}_{(l)}^{t_1}$, $\underline{m}_{(l)}^{t_1}$, $\overline{m}_{(l)}^{t_1}$, and $\overline{a}_{(l)}^{t_1}$ can be calculated for all attribute variables $\hat{x}^{t_1}$, $t_1=1,\ldots,m_1$, $l=1,\ldots,c$, using the formulae (4.19) – (4.26) and the corresponding fuzzy sets $B_l^{t_1}$ can be constructed. On the other hand, the technique used for the consequent learning is not changed for the case of interval-valued training data sets.

4.5.3 Illustrative Examples

The technique for constructing fuzzy inference systems will now be explained by examples. Firstly, the example for Sneath and Sokal's [104] two-dimensional data set will be considered. The data set is presented in Figure 2.8 and the clustering result obtained by using the D-AFC(c)-algorithm is presented in Figure 2.14.

The proposed technique for deriving fuzzy rules from fuzzy clusters was applied to the initial data. The membership functions $\gamma_{B_l^{t_1}}(\hat{x}^{t_1})$ and $\gamma_{C_l^l}(y_l)$ for the corresponding fuzzy sets $B_l^{t_1}$ and C_l^l, $t_1 = 1, 2$, $l = 1, 2$, were constructed immediately. The rule base induced by the clustering result obtained via the D-AFC(c)-algorithm can be seen in Figure 4.12 in which the performance of the designed fuzzy inference system is shown.

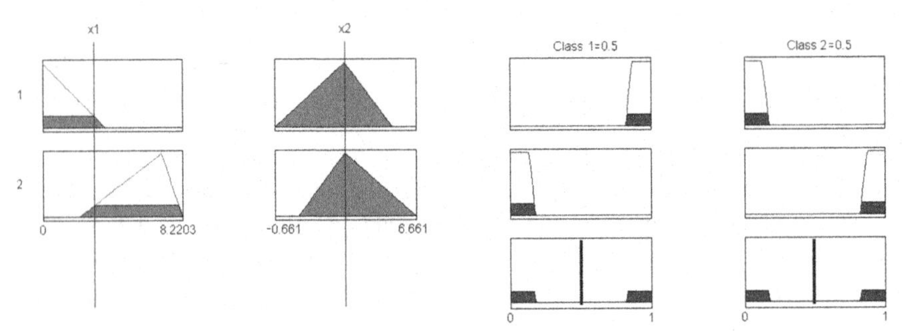

Fig. 4.12 Performance of the fuzzy inference system which was generated from Sneath and Sokal's [104] data set

Note that the fuzzy sets C_2^1 and C_1^2 are non-empty because the allotment $R_c^*(X)$, which corresponds to the clustering result, is the allotment among two particularly separated fuzzy clusters. Figure 4.12 shows an example of classification of the fifth object which belongs to both classes. Labels $x1$ and $x2$ denote the first attribute and the second attribute in the data set, and $l = 1, 2$ is the number of the rule in the figure.

The value of the membership function of the fuzzy cluster which corresponds to the first class is equal 0.92969 for the fifth object. The value of membership function of the second fuzzy cluster is equal 0.875 for the fifth object. So, $\mu_{15} > \mu_{25}$ and this fact is shown in Figure 2.14. On the other hand, Figure 4.12 shows that the fifth object is the element of both classes and the corresponding values of the average values of the range of output variables are equal 0.5. These values can be interpreted as the average values of belongingness of the object to both classes.

Let us consider the values of the average of the range of output variables which are the result of the defuzzification process. These values for all objects of Sneath and Sokal's [104] data set obtained from the generated fuzzy inference system are presented in Table 4.10.

Table 4.10 Values of the average of the range of output variables for Sneath and Sokal's [104] data set obtained from the fuzzy inference systems

Numbers of objects, i	Values of the average of the range of output variables		Numbers of objects, i	Values of the average of the range of output variables	
	Class 1	Class 2		Class 1	Class 2
1	0.919	0.0798	9	0.0824	0.919
2	0.924	0.0746	10	0.0824	0.919
3	0.915	0.0853	11	0.0853	0.915
4	0.918	0.0815	12	0.0808	0.920
5	0.500	0.5000	13	0.0764	0.925
6	0.918	0.0815	14	0.0808	0.920
7	0.918	0.0815	15	0.0853	0.915
8	0.915	0.0853	16	0.0853	0.915

Evidently, the results are correlated with the results obtained from the D-AFC(c)-algorithm so that it can be stated that the fuzzy inference system is accurate. On the other hand, the result which is obtained from the fuzzy inference system is easily interpretable. Thus, the obtained model is suitable for the interpretation since the rule consequents are the same or close to the actual class labels, so that each rule can be taken to describe all classes.

A plot of the output surface of the generated fuzzy inference system using both inputs and the second output is presented in Figure 4.13.

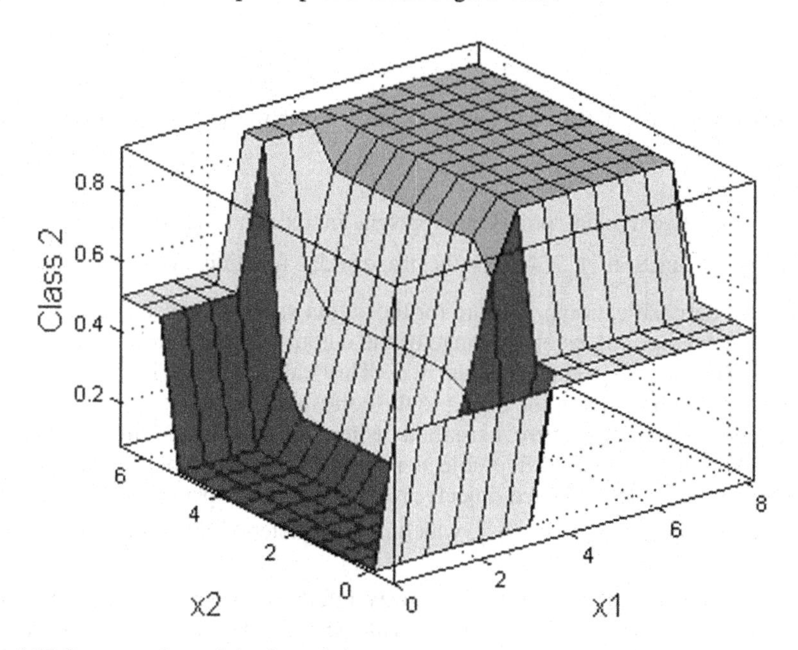

Fig. 4.13 Output surface of the fuzzy inference system

Secondly, let us consider a fuzzy inference system which was obtained from the clustering results of Anderson's Iris data set. The data were preprocessed according to the formulae (2.11), (2.15) and (1.26). The computational condition for the the D-AFC(c)-algorithm was determined as follows: the number of fuzzy clusters is $c = 3$ and the value of accuracy threshold is $\varepsilon = 0.0001$. The allotment $R_c^*(X)$ among three fully separated fuzzy clusters was obtained. The result of the experiment is considered in detail in the second chapter of the book.

The performance of the designed fuzzy inference system is shown in Figure 4.14. The corresponding rule base induced by the clustering result obtained by the D-AFC(c)-algorithm can be seen in this figure in which the labels SL l, SW l, PL l, and PW l denote, respectively, the sepal length, the sepal width, the petal length, and the petal width, and $l = 1,\ldots,3$ is the number of rule.

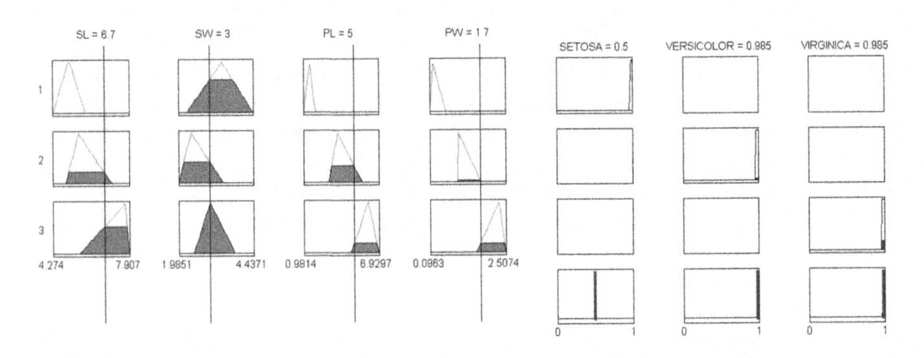

Fig. 4.14 Performance of the fuzzy inference system which was generated from Anderson's Iris data set

Note that two typical points are present in the first fuzzy cluster. So, the membership function $\gamma_{B_1^1}(\hat{x}^1)$ is the trapezoidal function. Moreover, the total area in the defuzzification procedure for the output variable Setosa is equal 0. That is why the average values of the range of the output variable Setosa is used as the output value and its value is equal 0.5. This value can be interpreted as an uncertain belongingness of the object to the corresponding class.

Anderson's Iris data were classified using the constructed fuzzy inference system. The rules classified five objects incorrectly and two objects were double-classified objects. The rejected objects are absent in the experiment. The example of classification of the object x_{147}, which is the double-classified one, is presented in Figure 4.14.

The application of the constructed fuzzy inference system to Anderson's Iris data was performed with a comparison with other approaches. Table 4.11 shows the results of such a comparison with some well-known classifiers.

Table 4.11 Comparison of results of different classifier systems on Anderson's Iris data set

Authors	The number of rules	The number of misclassifications
Höppner et al. [49]	8	6
Roubos et al. [96]	3	4
Ishibuchi et al. [55]	5	3
Abonyi et al. [2]	3	6
Abe et al. [1]	3	2

For example, Höppner, Klawonn, Kruse and Runkler [49] applied a simplified version of the GG-algorithm of fuzzy clustering to learn a Mamdani-type fuzzy inference system for classifying Anderson's Iris data by training on all 150 objects. An eight-rule fuzzy system was obtained. The rules classify 3 objects incorrectly and 3 more were not classified at all. So, the total number of misclassifications was 6.

On the other hand, the FCM-algorithm of fuzzy clustering was applied by Roubos and Setnes [96] to obtain an initial Takagi-Sugeno model with singleton consequents. All 150 samples were used in the training process. An initial model with three rules was constructed from clustering results where each rule described a class. The classification accuracy of the initial model was rather discouraging, giving 33 misclassifications on the training data. A multi-objective genetic algorithm-based optimization approach was applied to the initial model and as a result the number of misclassifications was reduced to 4.

Ishibuchi, Nakashima and Murata [55] applied all 150 samples in the training process, and derived a fuzzy classifier with five rules. The result was 3 misclassifications.

Abonyi, Roubos and Szeifert [2] proposed a data-driven method to design compact fuzzy classifiers via combining a genetic algorithm, a decision-tree initialization, and a similarity-driven rule reduction technique. The final fuzzy inference system had three fuzzy rules and the number of misclassifications was 6.

A fuzzy classifier with ellipsoidal regions was proposed by Abe and Thawonmas [1]. They applied clustering methods to extract fuzzy classification rules, with one rule around a cluster center, and then tuned the slops of the membership functions to obtain a high recognition rate. Finally, they obtained a three-rule fuzzy system with 2 misclassifications.

The results obtained from the fuzzy inference system proposed in his work are clearly at least comparable with the some well-known fuzzy systems.

Thirdly, let us consider an example of extracting fuzzy classification rules from the interval-valued training data set which was described in [143]. Sato and Jain's interval-valued data, which are presented in Table 3.8, were used in the experiment. The formula (3.51) and the function of dissimilarity (3.61) were used for data preprocessing. So, the membership functions of two classes of the allotment $R_c^*(X)$, which was obtained from the D-AFC(c)-algorithm, are as presented in Figure 3.23. The performance of the designed fuzzy inference system is shown in Figure 4.15.

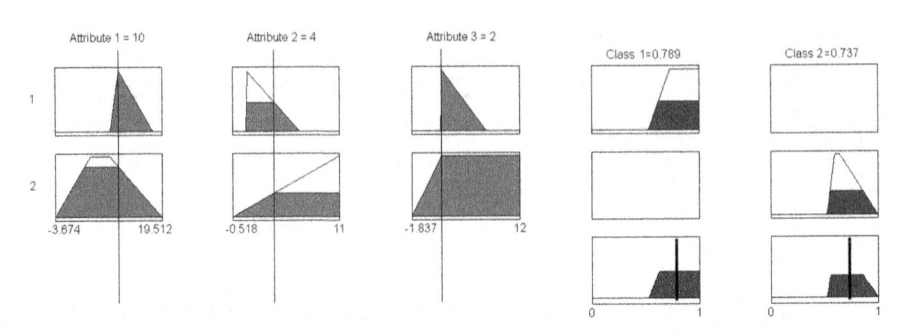

Fig. 4.15 Performance of the fuzzy inference system which was generated from Sato and Jain's interval-valued data set

The example of classification of some testing object $x_{test} = (\hat{x}^1_{test} = 10, \hat{x}^2_{test} = 4, \hat{x}^3_{test} = 2)$ is shown in Figure 4.15. Note that the values of the average of the range of output variables are different for both classes.

So, the results obtained using the proposed modeling approach for Sato and Jain's data set illustrate the effectiveness ad efficiency of the proposed method for the derivation of fuzzy classification rules from the interval-valued data. Obviously, the method can be generalized for extracting fuzzy rules from the three-way data.

The construction of a rule base from fuzzy clusters provides a first approximation for the data which can be used as a basis for further improvements. Some approaches, such as the genetic algorithm-based approach or neuro-fuzzy techniques, can be used for fuzzy rule tuning. Notably, the fuzzy rules obtained by using the D-AFC(c)-algorithm can be interpreted very simply because the membership functions of fuzzy sets which correspond to the input variables of fuzzy rules have a natural interpretation.

In general, the automatic method of designing fuzzy inference system for classification via a heuristic method of possibilistic clustering that has been described here can be considered as a suitable approach to rapid prototyping of fuzzy inference systems.

References

1. Abe, S., Thawonmas, R.: A fuzzy classifier with ellipsoidal regions. IEEE Transactions on Fuzzy Systems 5, 516–524 (1997)
2. Abonyi, J., Roubos, J.A., Szeifert, F.: Data-driven generation of compact, accurate and linguistically sound fuzzy classifiers based on a decision-tree initialization. International Journal of Approximate Reasoning 32, 1–21 (2003)
3. Anderson, E.: The irises of the Gaspe Peninsula. Bulletin of the American Iris Society 59, 2–5 (1935)
4. Atanassov, K.T.: Intuitionistic fuzzy sets. Fuzzy Sets and Systems 20, 87–96 (1986)
5. Atanassov, K.T.: Intuitionistic Fuzzy Sets: Theory and Applications. Springer, Heidelberg (1999)
6. Au, W.-H., Chan, K.C.C., Wong, A.K.C., Wang, Y.: Attribute clustering for grouping, selection and classification of gene expression data. IEEE/ACM Transactions on Computational Biology and Bioinformatics 2, 83–101 (2005)
7. Bandemer, H., Näther, W.: Fuzzy Data Analysis. Kluwer, Dordrecht (1992)
8. Bellmann, R.E., Kalaba, R., Zadeh, L.A.: Abstraction and pattern classification. Journal of Mathematical Analysis and Applications 13, 1–7 (1966)
9. Bensaid, A.M., Hall, L.O., Bezdek, J.C., Clarke, L.P.: Partially supervised clustering for image segmentation. Pattern Recognition 29, 859–871 (1996)
10. Bezdek, J.C.: Pattern Recognition with Fuzzy Objective Function Algorithms. Plenum Press, New York (1981)
11. Bezdek, J.C., Keller, J.M., Krishnapuram, R., Pal, N.R.: Fuzzy Models and Algorithms for Pattern Recognition and Image Processing. Springer Science, New York (2005)
12. Bouchachia, A., Pedrycz, W.: Enhancement of fuzzy clustering by mechanisms of partial supervision. Fuzzy Sets and Systems 157, 1733–1759 (2006)
13. Bouguessa, M., Wang, S., Sun, H.: An objective approach to cluster validation. Pattern Recognition Letters 27, 1419–1430 (2006)
14. Burillo, P., Bustince, H.: Intuitionistic fuzzy relations (Part I). Mathware and Soft Computing 2, 5–38 (1995)
15. Burillo, P., Bustince, H.: Intuitionistic fuzzy relations (Part II). Effect of Atanassov's operators on the properties of the intuitionistic fuzzy relations. Mathware and Soft Computing 2, 117–148 (1995)
16. Butkiewicz, B., Nieradka, G.: Fuzzy clustering for fuzzy features. In: De Baets, B., De Tré, G., Fodor, J., Kacprzyk, J., Zadrożny, S. (eds.) Current Issues in Data and Knowledge Engineering, pp. 102–107. EXIT, Warszawa (2004)
17. Cai, R., Lei, Y.J., Zhao, X.J.: Clustering method based on intuitionistic fuzzy equivalent dissimilarity matrix. Journal of Computer Applications 29, 123–126 (2009)
18. Chen, J.Y., Qin, Z., Jia, J.: A weighted mean subtractive clustering algorithm. Information Technology Journal 7, 356–360 (2008)

19. Chiang, J.-H., Yue, S., Yin, Z.-X.: A new fuzzy cover approach to clustering. IEEE Transactions on Fuzzy Systems 12, 199–208 (2004)
20. Chitsaz, E., Taheri, M., Katebi, S.D.: A fuzzy approach to clustering and selecting features for classification of gene expression data. In: Proceedings of the World Congress of Engineering, WCE 2008, London, United Kingdom, pp. 1650–1655. Newswood Limited, Hong Kong (2008)
21. Chitsaz, E., Taheri, M., Katebi, S.D., Jahromi, M.Z.: An improved fuzzy feature clustering and selection based on chi-squared-test. In: Proceedings of the International MultiConference of Engineers and Computer Scientists, IMECS 2009, pp. 35–40. Newswood Limited, Hong Kong (2009)
22. Chiu, S.L.: Fuzzy model identification based on cluster estimation. Journal of Intelligent and Fuzzy Systems 2, 267–278 (1994)
23. Coppi, R., D'Urso, P.: Fuzzy k-means clustering models for triangular fuzzy time trajectories. Statistical Methods and Applications 11, 21–40 (2002)
24. Coppi, R., D'Urso, P.: Three-way fuzzy clustering models for LR fuzzy time trajectories. Computational Statistics and Data Analysis 43, 149–177 (2003)
25. Corsini, P., Lazzerini, B., Marcelloni, F.: A new fuzzy relational clustering algorithm based on the fuzzy c-means algorithm. Soft Computing 9, 439–447 (2005)
26. Couturier, A., Fioleau, B.: Recognising stable corporate groups: a fuzzy classification method. Fuzzy Economic Review II, 35–45 (1997)
27. Damaratski, A., Novikau, D.: On the computational accuracy of the heuristic method of possibilistic clustering. In: Proceedings of the 10th International Conference on Pattern Recognition and Information Processing, PRIP 2009, pp. 78–81. Publishing Center BSU, Minsk (2009)
28. Damaratski, A., Juodelis, A.: A novel technique of partial supervision for a heuristic algorithm of possibilistic clustering. In: Proceedings of the 11th International Conference on Pattern Recognition and Information Processing, PRIP 2011, pp. 121–126. BSUIR, Minsk (2011)
29. Davé, R.N.: Use of the adaptive fuzzy clustering algorithm to detect lines in digital images. In: Casasent, D.P. (ed.) Intelligent Robots and Computer Vision VIII: Algorithms and Techniques: Proceedings of the Meeting Sponsored by SPIE, Philadelphia, USA, pp. 600–611. SPIE Press, Bellingham (1990)
30. De Cáceres, M., Oliva, F., Font, X.: On relational possibilistic clustering. Pattern Recognition 39, 2010–2024 (2006)
31. Devillez, A., Billaudel, P., Villermain Lecolier, G.: A fuzzy hybrid hierarchical clustering method with a new criterion able to find the optimal partition. Fuzzy Sets and Systems 128, 323–338 (2002)
32. Dong, Y., Zhuang, Y., Chen, K., Tai, X.: A hierarchical clustering algorithm based on fuzzy graph connectedness. Fuzzy Sets and Systems 157, 1760–1774 (2006)
33. Dubois, D., Prade, H.: Possibility Theory: An Approach to Computerized Processing of Uncertainty. Plenum Press, New York (1988)
34. Dumitrescu, D.: Hierarchical pattern classification. Fuzzy Sets and Systems 28, 145–162 (1988)
35. Dunn, J.C.: A fuzzy relative of the ISODATA process and its use in detecting compact well-separated clusters. Journal of Cybernetics 3, 32–57 (1974)
36. Friedman, M., Kandel, A.: Introduction to Pattern Recognition: Statistical, Structural, Neural and Fuzzy Logic Approaches. World Scientific, Singapore (1999)
37. Frigui, H., Krishnapuram, R.: Clustering by competitive agglomeration. Pattern Recognition 30, 1109–1119 (1997)
38. Frigui, H., Krishnapuram, R.: A robust algorithm for automatic extraction of an unknown number of clusters from noisy data. Pattern Recognition Letter 17, 1223–1232 (1996)

39. Gath, I., Geva, A.B.: Unsupervised optimal fuzzy clustering. IEEE Transactions on Pattern Analysis and Machines Intelligence 11, 773–780 (1989)
40. Geva, A.B.: Feature extraction and state identification in biomedical signals using hierarchical fuzzy clustering. Medical and Biological Engineering and Computing 36, 608–614 (1998)
41. Ghazavi, S.N., Liao, T.W.: Medical data mining by fuzzy modeling with selected features. Artificial Intelligence in Medicine 43, 195–206 (2008)
42. Gitman, J., Levine, M.D.: An algorithm for detecting unimodal fuzzy sets and its application as a clustering technique. IEEE Transactions on Computers 19, 583–593 (1970)
43. Grzegorzewski, P.: Distances between intuitionistic fuzzy sets and/or interval-valued fuzzy sets based on Hausdorff metric. Fuzzy Sets and Systems 148, 319–328 (2004)
44. Gustafson, D.E., Kessel, W.C.: Fuzzy clustering with a fuzzy covariance matrix. In: Gupta, M.M., Ragade, R.K., Yager, R.R. (eds.) Advances in Fuzzy Set Theory and Applications, pp. 605–620. North-Holland, Amsterdam (1979)
45. Hathaway, R.J., Davenport, J.W., Bezdek, J.C.: Relational duals of the c-means clustering algorithms. Pattern Recognition 22, 205–212 (1989)
46. Hathaway, R.J., Bezdek, J.C., Davenport, J.W.: On relational data versions of c-means algorithm. Pattern Recognition Letters 17, 607–612 (1996)
47. Hathaway, R.J., Bezdek, J.C.: NERF C-means: non-Euclidean relational fuzzy clustering. Pattern Recognition 27, 429–437 (1994)
48. Höppner, F.: Fuzzy shell clustering algorithms in image processing: fuzzy C-rectangular and 2-rectangular shells. IEEE Transactions on Fuzzy Systems 5, 599–613 (1997)
49. Höppner, F., Klawonn, F., Kruse, R., Runkler, T.: Fuzzy Cluster Analysis: Methods for Classification, Data Analysis and Image Recognition. Wiley, Chichester (1999)
50. Horng, Y.-J., Chen, S.-M., Chang, Y.-C., Lee, C.-H.: A new method for fuzzy information retrieval based on fuzzy hierarchical clustering and fuzzy inference techniques. IEEE Transactions on Fuzzy Systems 13, 216–238 (2005)
51. Huang, P.-F., Zhang, D.-Q.: Locality sensitive c-means clustering algorithms. Neurocomputing 73, 2935–2943 (2010)
52. Hung, W.-L., Lee, J.-S., Fuh, C.-D.: Fuzzy clustering based on intuitionistic fuzzy relations. International Journal of Uncertainty, Fuzziness and Knowledge-Based Systems 12, 513–529 (2004)
53. Hung, W.-L., Yang, M.-S.: Fuzzy clustering on LR-type fuzzy numbers with an application in Taiwanese tea evaluation. Fuzzy Sets and Systems 150, 561–577 (2005)
54. Iakovidis, D.K., Pelekis, N., Kotsifakos, E.E., Kopanakis, I.: Intuitionistic Fuzzy Clustering with Applications in Computer Vision. In: Blanc-Talon, J., Bourennane, S., Philips, W., Popescu, D., Scheunders, P. (eds.) ACIVS 2008. LNCS, vol. 5259, pp. 764–774. Springer, Heidelberg (2008)
55. Ishibuchi, H., Nakashima, T., Murata, T.: Three-objective genetic-based machine learning for linguistic rule extraction. Information Sciences 136, 109–133 (2001)
56. Ichino, M., Yaguchi, H.: Generalized Minkowski metrics for mixed feature-type data analysis. IEEE Transactions on Systems, Man, and Cybernetics 24, 698–708 (1994)
57. Ju, H., Yuan, X.: Similarity measures on interval-valued fuzzy sets and application to pattern recognition. In: Cao, D.Y. (ed.) Fuzzy Information and Engineering, pp. 875–883. Springer, Berlin (2007)
58. Kacprzyk, J.: Multistage Fuzzy Control. Wiley, Chichester (1997)
59. Karthikeyani Visalakshi, N., Thangavel, K., Parvathi, R.: An intuitionistic fuzzy approach to distributed fuzzy clustering. International Journal of Computer Theory and Engineering 2, 295–302 (2010)

60. Kaufmann, A.: Introduction to the Theory of Fuzzy Subsets. Academic Press, New York (1975)
61. Kaufman, L., Rousseeuw, P.J.: Finding Groups in Data: An Introduction to Cluster Analysis. Wiley, New York (1990)
62. Kaymak, U., Setnes, M.: Fuzzy clustering with volume prototypes and adaptive cluster merging. IEEE Transactions on Fuzzy Systems 10, 705–712 (2002)
63. Khan, H., Ahmad, M., Biswas, R.: Vague relations. International Journal of Computational Cognition 5, 31–35 (2007)
64. Kong, Y.-Q., Wang, S.-T.: Feature selection and semisupervised fuzzy clustering. Fuzzy Information and Engineering 2, 179–190 (2009)
65. Kreinovich, V., Kosheleva, O.: Towards dynamical systems approach to fuzzy clustering. In: Viattchenin, D.A. (ed.) Developments in Fuzzy Clustering, pp. 10–35. VEVER Publishing House, Minsk (2009)
66. Krishnapuram, R., Keller, J.M.: A possibilistic approach to clustering. IEEE Transactions on Fuzzy Systems 1, 98–110 (1993)
67. Krishnapuram, R., Keller, J.M.: The possibilistic c-means algorithm: insights and recommendations. IEEE Transactions on Fuzzy Systems 4, 385–393 (1996)
68. Krishnapuram, R., Joshi, A., Nasraoui, O., Yi, L.: Low-complexity fuzzy relational clustering algorithms for web mining. IEEE Transactions on Fuzzy Systems 9, 595–607 (2001)
69. Kuzmin, V.B.: A reference approach to obtaining fuzzy preference relations and the problem of choice. In: Yager, R.R. (ed.) Fuzzy Set and Possibility Theory: Recent Developments, pp. 107–118. Pergamon Press, New York (1982)
70. Kuzmin, V.B.: Constructing of Group Decisions in Spaces of Crisp and Fuzzy Binary Relations, Nauka, Moscow (1982) (in Russian)
71. Li, R., Mukaidono, M.: Gaussian clustering method based on maximum-fuzzy-entropy interpretation. Fuzzy Sets and Systems 102, 253–258 (1999)
72. Libert, G., Roubens, M.: Non-metric fuzzy clustering algorithms and their cluster validity. In: Gupta, M.M., Sanchez, E. (eds.) Approximate Reasoning in Decision Analysis, pp. 417–425. North-Holland, Amsterdam (1982)
73. Libert, G.: Compactness and number of clusters. Control and Cybernetics 15, 205–212 (1986)
74. Łęski, J.M.: Robust possibilistic clustering. Archives of Control Sciences 10, 141–155 (2000)
75. Mamdani, E.H., Assilian, S.: An experiment in linguistic synthesis with a fuzzy logic controller. International Journal of Man-Machine Studies 7, 1–13 (1975)
76. Mandel, I.D.: Clustering Analysis. Finansy i Statistica, Moscow (1988) (in Russian)
77. Ménard, M., Courboulay, V., Dardignac, P.-A.: Possibilistic and probabilistic fuzzy clustering: unification within the framework of the non-extensive thermostatistics. Pattern Recognition 36, 1325–1342 (2003)
78. Mirkin, B.G.: Analysis of Qualitative Attributes and Structures, Statistika, Moscow (1980) (in Russian)
79. Miyamoto, S., Mukaidono, M.: Fuzzy c-means as a regularization and maximum entropy approach. In: Mares, M., Mesiar, R., Novak, V., Ramik, J., Stupnanova, A. (eds.) Proceedings of the 7th International Fuzzy Systems Association World Congress, IFSA 1997, vol. 1, pp. 86–97. Academia, Prague (1997)
80. Miyamoto, S., Ichihashi, H., Honda, K.: Algorithms for Fuzzy Clustering Methods in C-Means Clustering with Applications. Springer, Heidelberg (2008)
81. Nojiri, H.: A model of fuzzy team decision. Fuzzy Sets and Systems 2, 201–212 (1979)
82. Orlovsky, S.A.: Problems of Decision Making for Fuzzy Initial Information, Nauka, Moscow (1981) (in Russian)

83. Ovchinnikov, S.: Similarity relations, fuzzy partitions, and fuzzy orderings. Fuzzy Sets and Systems 40, 107–126 (1991)
84. Owsiński, J.W.: Asymmetric distances – a natural case for fuzzy clustering? In: Viattchenin, D.A. (ed.) Developments in Fuzzy Clustering, pp. 36–45. VEVER Publishing House, Minsk (2009)
85. Pal, N.R., Pal, K., Bezdek, J.C.: A mixed c-means clustering model. In: Proceedings of the 6th IEEE International Conference on Fuzzy Systems, FUZZ-IEEE 1997, Barcelona, Spain, pp. 11–21. IEEE Service Center, Piscataway (1997)
86. Pal, N.R., Pal, K., Keller, J.M., Bezdek, J.C.: A possibilistic fuzzy c-means clustering algorithm. IEEE Transactions on Fuzzy Systems 13, 517–530 (2005)
87. Pal, N.R., Chakraborty, D.: Mountain and subtractive clustering method: improvements and generalizations. International Journal of Intelligent Systems 15, 329–341 (2000)
88. Pedrycz, W.: Algorithms of fuzzy clustering with partial supervision. Pattern Recognition Letters 3, 13–20 (1985)
89. Pedrycz, W.: Fuzzy sets in pattern recognition: methodology and methods. Pattern Recognition 23, 121–146 (1990)
90. Pedrycz, W.: Conditional fuzzy c-means. Pattern Recognition Letters 17, 625–631 (1996)
91. Pedrycz, W., Waletzky, J.: Fuzzy clustering with partial supervision. IEEE Transactions on Systems, Man, and Cybernetics – Part B: Cybernetics 27, 787–795 (1997)
92. Pedrycz, A., Reformat, M.: Hierarchical FCM in a stepwise discovery of structure in data. Soft Computing 10, 244–256 (2006)
93. Pelekis, N., Iakovidis, D.K., Kotsifakos, E.E., Kopanakis, I.: Fuzzy clustering of intuitionistic fuzzy data. International Journal of Business Intelligence and Data Mining 3, 45–65 (2008)
94. Radecki, T.: Level fuzzy sets. Journal of Cybernetics 7, 189–198 (1977)
95. Roubens, M.: Pattern classification problems and fuzzy sets. Fuzzy Sets and Systems 1, 239–253 (1978)
96. Roubos, H., Setnes, M.: Compact and transparent fuzzy models and classifiers through iterative complexity reduction. IEEE Transactions on Fuzzy Systems 9, 516–524 (2001)
97. Ruspini, E.H.: A new approach to clustering. Information and Control 15, 22–32 (1969)
98. Ruspini, E.H.: Numerical methods for fuzzy clustering. Information Sciences 2, 319–350 (1970)
99. Ruspini, E.H.: Recent developments in fuzzy clustering. In: Yager, R.R. (ed.) Fuzzy Set and Possibility Theory: Recent Developments, pp. 133–146. Pergamon Press, New York (1982)
100. Sato, M., Sato, Y.: On a multicriteria fuzzy clustering method for 3-way data. International Journal of Uncertainty, Fuzziness and Knowledge-Based Systems 2, 127–142 (1994)
101. Sato, M., Sato, Y., Jain, L.C.: Fuzzy Clustering Models and Applications. Springer, Heidelberg (1997)
102. Sato-Ilic, M., Jain, L.C.: Innovations in Fuzzy Clustering: Theory and Applications. Springer, Heidelberg (2006)
103. Sledge, I.J., Havens, T.C., Bezdek, J.C., Keller, J.M.: Relational cluster validity. In: Aranda, J., Xambó, S. (eds.) 2010 IEEE World Congress on Computational Intelligence, WCCI 2010, Barcelona, Spain. Plenary and Invited Lectures, pp. 151–170. IEEE Service Center, Piscataway (2010)
104. Sneath, P.H.A., Sokal, R.: Numerical Taxonomy. Freeman, San Francisco (1973)

105. Szmidt, E., Kacprzyk, J.: Classification with Nominal Data Using Intuitionistic Fuzzy Sets. In: Melin, P., Castillo, O., Aguilar, L.T., Kacprzyk, J., Pedrycz, W. (eds.) IFSA 2007. LNCS (LNAI), vol. 4529, pp. 76–85. Springer, Heidelberg (2007)
106. Tamura, S., Higuchi, S., Tanaka, K.: Pattern classification based on fuzzy relations. IEEE Transactions on Systems, Man, and Cybernetics 1, 61–66 (1971)
107. Todorova, L., Vassilev, P.: Algorithm for clustering data set represented by intuitionistic fuzzy estimates. International Journal of Bioautomation 14, 61–68 (2010)
108. Torra, V., Miyamoto, S., Endo, Y., Domingo-Ferrer, J.: On intuitionistic fuzzy clustering for its application to privacy. In: Proceedings of the 2008 IEEE World Congress on Computational Intelligence (WCCI 2008) and the 17th IEEE International Conference on Fuzzy Systems (FUZZ-IEEE 2008), Hong Kong, China, pp. 1042–1048. IEEE Service Center, Piscataway (2008)
109. Vapnik, V.N.: Statistical Learning Theory. Wiley, New York (1998)
110. Viattchenin, D.A.: On projections of fuzzy similarity relations. In: Proceedings of the 5th International Conference on Computer Data Analysis and Modeling, CDAM 1998, Minsk, Belarus, vol. 2, pp. 150–155. Publishing Center BSU, Minsk (1998)
111. Viattchenin, D.A.: Human-computer approach to fuzzy classification problem based on the concept of representation. In: Proceedings of the 2nd International Conference on Quality of Life Statistical Data Analysis, QoL 2002, Wroclaw, Poland, pp. 189–204. Wroclaw University of Economics, Wrocław (2002)
112. Viattchenin, D.A.: Criteria of quality of allotment in fuzzy clustering. In: Proceedings of the 3th International Conference on Neural Networks and Artificial Intelligence, ICNNAI 2003, Minsk, Belarus, pp. 91–94. Belarusian State University of Informatics and Radioelectronics, Minsk (2008)
113. Viattchenin, D.A.: Fuzzy Methods of Automatic Classification. Technoprint Publishing House, Minsk (2004) (in Russian)
114. Viattchenin, D.A.: A new heuristic algorithm of fuzzy clustering. Control and Cybernetics 33, 323–340 (2004)
115. Viattchenin, D.A.: On the number of fuzzy clusters in the allotment. In: Proceedings of the 7th International Conference on Computer Data Analysis and Modeling, CDAM 2004, vol. 1, pp. 198–201. Publishing Center BSU, Minsk (2004)
116. Viattchenin, D.A.: Parameters of the AFC-method of fuzzy clustering. Bulletin of The Military Academy of The Republic of Belarus 4(5), 51–55 (2004) (in Russian)
117. Viattchenin, D.A.: Level fuzzy relations and their applications in pattern recognition. Bulletin of The Military Academy of The Republic of Belarus 4(9), 25–31 (2005) (in Russian)
118. Viattchenin, D.A.: Fast algorithms of the decomposition of fuzzy tolerances and its modifications for construction of initial allotments in the AFC–method of fuzzy clustering. Proceedings of The Military Academy of The Republic of Belarus 8, 62–67 (2005) (in Russian)
119. Viattchenin, D.A., Savyhin, P., Sharamet, A.: Hierarchical AFC-algorithm of fuzzy clustering based on the transitive closure operation. Proceedings of The Military Academy of The Republic of Belarus 10, 39–48 (2006) (in Russian)
120. Viattchenin, D.A.: On the inspection of classification results in the fuzzy clustering method based on the allotment concept. In: Proceedings of the 4th International Conference on Neural Networks and Artificial Intelligence, ICNNAI 2006, pp. 210–216. Brest State Technical University, Brest (2006)
121. Viattchenin, D.A.: Distances between type 2 fuzzy sets and their applications to solving of identification problems. Bulletin of The Military Academy of The Republic of Belarus 3(12), 11–17 (2006) (in Russian)

122. Viattchenin, D.A.: Heuristics for detection of an allotment among unknown number of fuzzy clusters. In: Proceedings of the 9th International Conference on Pattern Recognition and Information Processing, PRIP 2007, vol. II, pp. 220–225. United Institute of Informatics Problems of National Academy of Sciences of Belarus, Minsk (2007)
123. Viattchenin, D.A.: A direct algorithm of possibilistic clustering with partial supervision. Journal of Automation, Mobile Robotics and Intelligent Systems 1(3), 29–38 (2007)
124. Viattchenin, D.A.: A methodology of fuzzy clustering with partial supervision. Systems Sciences 33(4), 61–71 (2007)
125. Viattchenin, D.A.: A proximity-based fuzzy clustering of fuzzy numbers. In: Proceedings of the 8th International Conference on Computer Data Analysis and Modeling, CDAM 2007, vol. 1, pp. 182–185. Publishing Center BSU, Minsk (2007)
126. Viattchenin, D.A.: Direct algorithms of fuzzy clustering based on the transitive closure operation and their application to outliers detection. Artificial Intelligence 3, 205–216 (2007) (in Russian)
127. Viattchenin, D.A.: Fuzzy objective function-based technique of partial supervision for a heuristic method of possibilistic clustering. In: Proceedings of the 5th International Conference on Neural Networks and Artificial Intelligence, ICNNAI 2008, pp. 51–55. PROPILEI Publishing House, Minsk (2008)
128. Viattchenin, D.A.: Method of soft interpretation of fuzzy clustering results. Taurian Herald for Computer Science and Mathematics 1, 107–114 (2008) (in Russian)
129. Viattchenin, D.A.: On possibilistic interpretation of membership values in fuzzy clustering method based on the allotment concept. Proceedings of the Institute of Modern Knowledge 3(36), 85–90 (2008) (in Russian)
130. Viattchenin, D.A.: Notes on the decomposition of intuitionistic fuzzy relations. In: Proceedings of the 7th International Workshop on Intuitionistic Fuzzy Sets and Generalized Nets, IWIFSGN 2008, vol. I, pp. 251–262. EXIT, Warsaw (2008)
131. Viattchenin, D.A.: Discriminating fuzzy preference relations based on heuristic possibilistic clustering. In: Owsiński, J.W., Brüggemann, R. (eds.) Multicriteria Ordering and Ranking: Partial Orders, Ambiguities and Applied Issues, pp. 197–213. Systems Research Institute, Polish Academy of Sciences, Warsaw (2008)
132. Viattchenin, D.A.: A heuristic approach to possibilistic clustering for fuzzy data. Journal of Information and Organizational Sciences 32, 149–163 (2008)
133. Viattchenin, D.A.: Kinds of fuzzy α-clusters. Proceedings of the Institute of Modern Knowledge 4(37), 95–101 (2008) (in Russian)
134. Viattchenin, D.A.: An outline for a heuristic approach to possibilistic clustering of the three-way data. Journal of Uncertain Systems 3, 64–80 (2009)
135. Viattchenin, D.A.: An algorithm for detecting the principal allotment among fuzzy clusters and its application as a technique of reduction of analyzed features space dimensionality. Journal of Information and Organizational Sciences 33, 205–217 (2009)
136. Viattchenin, D.A., Damaratski, A., Novikau, D.: Relational clustering of heterogeneous fuzzy data. In: Viattchenin, D.A. (ed.) Developments in Fuzzy Clustering, pp. 74–91. VEVER Publishing House, Minsk (2009)
137. Viattchenin, D.A.: Analysis of the cluster structure robustness in non-stationary clustering problems. Doklady BGUIR 6(44), 91–98 (2009) (in Russian)
138. Viattchenin, D.A.: Partial training method for heuristic algorithm of possibilistic clustering under unknown number of classes. Vestnik BNTU 5, 67–74 (2009) (in Russian)

139. Viattchenin, D.A., Damaratski, A.: Constructing of allotment among fuzzy clusters in case of quasi-robust cluster structure of set of objects. Doklady BGUIR 1(47), 46–52 (2010) (in Russian)
140. Viattchenin, D.A.: An outline for a new approach to clustering based on intuitionistic fuzzy relations. Notes on Intuitionistic Fuzzy Sets 16, 40–60 (2010)
141. Viattchenin, D.A., Klawonn, F., Tschumitschew, K.: A new validity measure for heuristic possibilistic clustering. In: Proceedings of the 6th International Conference on Neural Networks and Artificial Intelligence, ICNNAI 2010, pp. 74–79. Brest State Technical University, Brest (2010)
142. Viattchenin, D.A.: Automatic generation of fuzzy inference systems using heuristic possibilistic clustering. Journal of Automation, Mobile Robotics and Intelligent Systems 4(3), 36–44 (2010)
143. Viattchenin, D.A.: Derivation of fuzzy rules from interval-valued data. International Journal of Computer Applications 7(3), 13–20 (2010)
144. Viattchenin, D.A.: Validity measures for heuristic possibilistic clustering. Information Technology and Control 39, 321–332 (2010)
145. Viattchenin, D.A.: A heuristic possibilistic approach to clustering for asymmetric data. Journal of Applied Computer Science and Mathematics 10(5), 87–92 (2011)
146. Viattchenin, D.A.: An approach to constructing a possibility distribution for the number of fuzzy clusters. In: Proceedings of the 11th International Conference on Pattern Recognition and Information Processing, PRIP 2011, pp. 188–194. BSUIR, Minsk (2011)
147. Viattchenin, D.A., Damaratski, A.: Constructing fuzzy c-partition in case of unstable cluster structure of set of objects. Artificial Intelligence 3, 479–489 (2011) (in Russian)
148. Viattchenin, D.A.: Constructing stable clustering structure for uncertain data set. Acta Electrotechnica et Informatica 11(3), 42–50 (2011)
149. Viertl, R.: Fuzzy data and statistical modeling. In: Proceedings of the 8th International Conference on Computer Data Analysis and Modeling, CDAM 2007, vol. 1, pp. 108–114. Publishing Center BSU, Minsk (2007)
150. Vlachos, I.K., Sergiadis, G.D.: Intuitionistic fuzzy information – applications to pattern recognition. Pattern Recognition Letters 28, 197–206 (2006)
151. Walesiak, M.: Ugólniona miara odległości w statystycznej analizie wielowymiarowej. Wydawnictwo Akademii Ekonomicznej im. Oskara Langego, Wrocław (2002) (in Polish)
152. Wang, S.T., Chung, F.L., Deng, Z., Hu, D., Wu, X.: Robust maximum entropy clustering algorithm with its labeling for outliers. Soft Computing 10, 555–563 (2006)
153. Wang, Z., Xu, Z., Liu, S., Tang, J.: A netting clustering analysis method under intuitionistic fuzzy environment. Applied Soft Computing 11, 5558–5564 (2011)
154. Watada, J., Tanaka, H., Asai, K.: Recent developments in fuzzy clustering. In: Yager, R.R. (ed.) Fuzzy Set and Possibility Theory: Recent Developments, pp. 148–166. Pergamon Press, New York (1982)
155. Windham, M.P.: Numerical classification of proximity data with assignment measures. Journal of Classification 2, 157–172 (1985)
156. Wu, K.-L., Yang, M.-S.: Alternative c-means clustering algorithms. Pattern Recognition 35, 2267–2278 (2002)
157. Wu, X., Wu, B., Sun, J., Fu, H.: Unsupervised possibilistic fuzzy clustering. Journal of Information and Computational Science 7, 1075–1080 (2010)
158. Xie, X.L., Beni, G.: A validity measure for fuzzy clustering. IEEE Transactions on Pattern Analysis and Machines Intelligence 13, 841–847 (1991)

159. Xie, Z., Wang, S.T., Chung, F.L.: An enhanced possibilistic c-means clustering algorithm EPCM. Soft Computing 12, 593–611 (2008)
160. Xu, Z., Chen, J., Wu, J.: Clustering algorithm for intuitionistic fuzzy sets. Information Sciences 178, 3775–3790 (2008)
161. Xu, Z.: Intuitionistic fuzzy hierarchical clustering algorithms. Journal of Systems Engineering and Electronics 20, 1–8 (2009)
162. Xu, Z., Wu, J.: Intuitionistic fuzzy c-means clustering algorithms. Journal of Systems Engineering and Electronics 21, 580–590 (2010)
163. Yager, R.R., Filev, D.P.: Approximate clustering via the mountain method. IEEE Transactions on Systems, Man, and Cybernetics 24, 1279–1284 (1994)
164. Yangand, M.-S., Ko, C.-H.: On a class of fuzzy c-numbers clustering procedures for fuzzy data. Fuzzy Sets and Systems 84, 49–60 (1996)
165. Yang, M.-S., Liu, H.-H.: Fuzzy clustering procedures for conical fuzzy vector data. Fuzzy Sets and Systems 106, 189–200 (1999)
166. Yang, M.-S., Shih, H.-M.: Cluster analysis based on fuzzy relations. Fuzzy Sets and Systems 120, 197–212 (2001)
167. Yang, M.-S., Wu, K.-L.: A modified mountain clustering algorithm. Pattern Analysis and Applications 8, 125–138 (2005)
168. Yang, M.-S., Wu, K.-L.: Unsupervised possibilistic clustering. Pattern Recognition 39, 5–21 (2006)
169. Zadeh, L.A.: Fuzzy sets. Information and Control 8, 338–353 (1965)
170. Zadeh, L.A.: Similarity relations and fuzzy orderings. Information Sciences 3, 177–200 (1971)
171. Zadeh, L.A.: The Concept of a Linguistic Variable and Its Application to Approximate Reasoning. American Elsevier Publishing Company, New York (1973)
172. Zadeh, L.A.: Fuzzy sets as the basis for a theory of possibility. Fuzzy Sets and Systems 1, 3–28 (1978)
173. Zeng, J., Liu, Z.-Q.: Type-2 fuzzy sets for pattern recognition: the state-of-the-art. Journal of Uncertain Systems 1, 163–177 (2007)
174. Zhang, J.-S., Leung, Y.-W.: Improved possibilistic c-means clustering algorithms. IEEE Transactions on Fuzzy Systems 12, 209–217 (2004)
175. Zimmermann, H.-J.: Fuzzy Set Theory and Its Applications. Kluwer, Boston (1991)
176. Atanassov, K.T.: On Intuitionistic Fuzzy Sets Theory. Springer, Heidelberg (2012)

Printed in the United States
By Bookmasters